A CATHOLIC CASE
FOR
INTELLIGENT DESIGN

A Catholic Case
for
Intelligent Design

Fr. Martin Hilbert, C.O.

Seattle   Discovery Institute Press   2024

Description

A faithful catechist in Fr. Martin Hilbert's parish came to see him. "Father Martin," she said, "I have been teaching children about Adam and Eve, just as the *Catechism* tells us. But we can't be expected to believe that, can we? What is the real story?" Her question was the catalyst for *A Catholic Case for Intelligent Design*. In taut, accessible prose, Fr. Hilbert draws upon his broad learning in science, philosophy, history, and theology to show that modern evolutionary theory, including theistic evolution, faces a rising wave of disconfirming evidence. Meanwhile, the evidence for both intelligent design and a first human couple, Adam and Eve, is stronger than ever. What about the problem of suffering, disease, and death in a world created by a wise and good Creator? Fr. Hilbert tackles that issue as well, and explains why the theory of intelligent design, rightly understood, harmonizes perfectly with the Catholic theological tradition.

Copyright Notice

© 2024 by Discovery Institute. All Rights Reserved.

Library Cataloging Data

A Catholic Case for Intelligent Design by Fr. Martin Hilbert, C.O.

Cover design by Tri Widyatmaka.

346 pages, 6 x 9 inches

Library of Congress Control Number: 2024949219

ISBN: 978-1-63712-071-2 (paperback), 978-1-63712-073-6 (Kindle), 978-1-63712-072-9 (EPUB)

BISAC: REL106000   RELIGION / Religion & Science

BISAC: SCI027000   SCIENCE / Life Sciences / Evolution

BISAC: REL010000   RELIGION / Christianity / Catholic

Publisher Information

Discovery Institute Press, 208 Columbia Street, Seattle, WA 98104

Internet: discovery.press

Published in the United States of America on acid-free paper.

First Edition, November 2024

Advance Praise

In 1950 Pope Pius XII stated that discussion regarding the hypothesis of the evolutionary origin of the human body is not forbidden for Catholic theologians. However, the expected debate never happened. Theologians, by and large, interpreted the Pope's permission as unequivocal support for an essentially naturalistic account of human origins. Now, over seventy years later, the theological community finally has a chance to take the turn indicated by the Pope and ask fundamental questions about modern evolutionary theory, intelligent design, human origins, and the Christian understanding of creation. Fr. Hilbert's book is an important voice in this debate. He does not follow the easy and greatly wanting path of mixing Christianity with naturalism in the form of theistic evolution. Instead, he looks at evidence with an open mind and incorporates anything good and true he finds in modern science and traditional theology. This book will be an eye-opener for those who never thought that a Catholic can support intelligent design and be scientifically informed.
—**Fr. Michael Chaberek**, PhD, member of the Polish Dominican Province; author of *Catholicism and Evolution* (2015) and *Aquinas and Evolution* (2017)

For too long advocates of Darwinism have taken advantage of the openness of the Catholic Church to true science in order to insinuate their barely disguised materialistic philosophy. In easy-to-follow writing, Fr. Martin Hilbert summarizes why modern scientific evidence stacks decidedly against Darwinism—and strongly in favor of the

traditional Catholic understanding that life was purposely made by an intelligent designer.
—**Michael J. Behe**, PhD, professor of biological sciences, Lehigh University; author of *Darwin's Black Box*, *The Edge of Evolution*, and *Darwin Devolves*

What a delightful discovery this book has been, superb not only from a theological and philosophical point of view, but also by biological, biochemical, anthropological, and even engineering standards. In crystal clear prose, Hilbert shows that Darwinism is not so much the most successful contemporary *theory* of human origins as the most powerful contemporary *myth* of human origins, one which slants our views of who we are and how we are related to our Designer in unwarranted ways. Naturally his defense of the classical Christian view will deeply interest Catholics, but by rights his book should command the attention of all people interested in thinking clearly.
—**J. Budziszewski**, Professor of Government and Philosophy, University of Texas at Austin, author of *Commentary on Thomas Aquinas's Treatise on the One God* (Cambridge University Press)

Fr. Hilbert has written a deeply insightful and broad defense of intelligent design for Catholics, based on scientific, philosophical, and theological arguments. He thoughtfully distinguishes claims that are clearly supported by the evidence from questions difficult to answer conclusively. The book is also highly engaging as he shares his own personal journey of seeking the truth and explores the intrigue of individuals and social forces that has led many Christians into materialist intellectual captivity. This work is a must-read for anyone concerned about the future of the Church.
—**Brian Miller**, PhD, Research Coordinator and Senior Fellow for the Center for Science and Culture, an organizer of the Conference on Engineering in the Life Sciences (CELS), and contributor to multiple books and journals covering the debate over intelligent design, including *The Mystery of Life's Origin: The Continuing Controversy* and *Inference Review*

"Children by nature are theists." So begins *A Catholic Case for Intelligent Design* by Fr. Martin Hilbert, who as a child escaped Communism, but not before he "knew that evil was real." Later he came to the view that Darwinism, whatever else it might be, "was an attack on the power of reason to know reality and to arrive at the Creator." He holds a PhD in the history and philosophy of science and teaches a course on the subject to seminarians. But he is by no means an ivory tower intellectual. He holds a master's in electrical engineering and pastors a Catholic parish, dealing with the day-to-day concerns and suffering of his flock. Fr. Hilbert's good sense shines through every page of this brave and refreshing work.

As a Catholic physician, I couldn't agree more with what this highly educated, well-rounded, down-to-earth spiritual father thinks of Darwinism. "Its primary purpose is to serve as a creation myth for the secular society in which we live," he writes. "In this role it contributes to an impoverished view of humanity and society, and is a constant threat to the Christian understanding of the creation and fall of man. Moreover, the thought patterns it engenders are destructive of common sense and of a responsible use of reason. It is, I contend, an unmitigated intellectual disaster, whose myriad shortcomings need to be dragged into broad daylight." Indeed! And that's exactly what this book accomplishes. It's a comprehensive, up-to-date exposure and take-down of Darwinism.

Fr. Hilbert also provides a nuanced look at the theory of intelligent design, clearing up common misconceptions about the research program and critically analyzing the explanations and key claims of its leading proponents, and doing so in the light of perennial philosophy and the teachings of the Catholic Church.

For Catholics who find their faith undermined by prominent spokespersons for contemporary science, Fr. Hilbert has a message of hope, one that draws an important distinction between actual scientific evidence on the one hand, and on the other hand, scientism and materialism masquerading as science. "The good news for the Catholic is that the contemporary scientific worldview is false," he writes. "If the world at large has not heard of Darwinism's demise, it is because the work of the ID community has been dismissed and ridiculed, and a collection of fawning Catholics and other Christians with prominent platforms have rushed to the defense of Darwinism and tried to change the Church's teachings to accommodate it. Yet in the midst of this sad state of affairs, the faithful need not worry that the Church has been wrong on the origin of life, man, and sin."

One senses that Fr. Hilbert was driven to write this incisive and wide-ranging book. As a kindred spirit, who hails from his alma mater, I am sure that when the time comes, he will hear what all of us, deep in our hearts, hope to hear: "Well done my good and faithful servant; come share your master's joy" (Matthew 25:21).

—**Howard Glicksman**, MD, co-author of *Your Designed Body*

Faithful Catholics should be delighted with the emerging new evidence for intelligent design in nature. The evidence confirms the settled Catholic teaching that nature is teleological and that we can know by reason from the creation that a Creator exists. But for obscure reasons, many Catholic academics object to intelligent design. Fr. Martin Hilbert is the perfect person to dispel these objections. His knowledge of theology, philosophy, engineering, and the philosophy of science allows him to navigate the troubled waters of the intelligent design debates deftly and accessibly.

I'm especially fond of the way Fr. Hilbert integrates traditional philosophical arguments for God's existence with the more empirically focused—and modest—arguments for intelligent design. *A Catholic Case for Intelligent Design* is bound to become a classic contribution to the growing literature on ID. Bravo!

—**Jay W. Richards**, editor of (and contributor to) *God and Evolution: Protestants, Catholics, and Jews Explore Darwin's Challenge to Faith*, co-author of *The Privileged Planet: How Our Place in the Cosmos Is Designed for Discovery*, and Director of the Richard and Helen DeVos Center for Life, Religion, and Family at The Heritage Foundation

*To the Fathers and Brothers of the
Toronto Oratory of Saint Philip Neri
and in memory of Fr. Jonathan Robinson
and of my parents, Peter and Ludmila Hilbert*

Contents

1. Introduction: Wrestling with Darwin 13
2. Evolution: More than a Hypothesis 33
3. Intelligent Design in Nature 93
4. Intelligent Design: A Preamble to a Powerful Way to God .. 141
5. Creation Groans 161
6. Prehistoric Man 177
7. Man: The Image of God 211
8. Anti-Theist Darwin and His Useful Instruments 241
9. Converging and Convincing Arguments 273
   Appendix A. Some Notes on Transformism 289
   Appendix B. When Did Adam and Eve Live? 295
   Appendix C. Faith of Our Fathers—
      A Hermeneutic of Continuity 299
   Endnotes 303
   Figure Credits 335
   Index 337

1. Introduction: Wrestling with Darwin

> *I have lately read Morley's* Life of Voltaire *& he insists strongly that direct attacks on Christianity (even when written with the wonderful force & vigour of Voltaire) produce little permanent effect: real good seems only to follow from slow & silent side attacks.*
> —Charles Darwin, letter to his son George[1]

Children are by nature theists, with an inherent sense that things and events have causes, that the world is a wonderful place, and that wonderful things call for wonderful causes. In this I was no different.

I was ten or eleven when I first encountered Darwin's alternative theory. My first thought was that the whole business was an attack on God. It was not that I was wedded to a naively literal interpretation of the book of Genesis. Rather, I perceived that Darwinism was an attack on the power of reason to know reality and to arrive at the Creator. Animals give birth to their own kind. Surely everyone knows that, I thought. In my youthful eyes, Darwin's theory was something wicked people had devised to dispense with God and the Ten Commandments. Having escaped from Communist Czechoslovakia in 1968, I knew that evil was real; and it seemed to me that this theory was somehow entangled with it.

My youthful perspective lacked all nuance, to be sure; and yet later I was to discover that both the founders of modern evolutionary theory and some of its leading contemporary proponents had explicitly confessed to motives bracingly akin to my cartoon-like sense of their reasons for embracing the theory.

Such discoveries, however, were far in the future. After my parents assured me that the theory was far from proven, I did not give it much thought. In high school and college, I focused on math and physics and eventually graduated from university with a master's degree in electrical engineering.

It was in graduate school that I began to take a greater interest in my Catholic faith. I had a lot of catching up to do, because the last time I'd had any formal instruction in religion was a few sessions in grade eight, in preparation for confirmation. So when I wasn't solving Maxwell's equations and the like, I was reading C. S. Lewis, Fulton Sheen, Thomas Merton, G. K. Chesterton, and various catechisms.

They felt like two separate worlds to me, but on one occasion, the two passions came together. When reading J. D. Jackson's *Classical Electrodynamics*, I encountered a graph of the translucence of water as a function of frequency. The graph showed that in the extremely narrow range of optical frequencies, water becomes translucent, whereas at most frequencies below and above this "window," it is nearly opaque. (The difference is many orders of magnitude.) Here was an instance of extreme fine-tuning of chemistry to allow for vision. The fact jumped out of the page at me. If it were not for this "window" existing and being situated precisely where it is, no animal could see, and photosynthesis could not take place. One possible explanation—and indeed, the most obvious one—seemed to leap off the page: This exquisite fine-tuning required an exquisitely skilled fine-tuner, a designing intelligence at the very foundations of the molecular and atomic order of nature.

The graph was too technical to explain to most people. And the ones who understood it—my engineering classmates—dismissed it as Jackson had tried to do himself; in their minds, it just showed the power of natural selection to design the eye around this optical window. The graph did not convert any of my fellow grad students to belief in God.

Immediately upon finishing my engineering degree, I joined the Oratory of Saint Philip Neri; and, five years later, I was ordained a priest. I never lost my fascination with science, and I always retained a confidence that it could reveal important truths about the world. One day, at some point around 1990, I came upon an article by George Sim Johnston in the Oratory library,[2] in which he criticized Darwinism for its lack of empirical evidence. I had come across some of these arguments before, but what was new to me was the smoking-gun evidence that Charles Darwin intentionally sought to rid the world of Christianity.

Johnston quoted a letter written by Charles Darwin to his son George in 1873, part of which I quoted at the top of this chapter, but it bears repeating: "I have lately read Morley's *Life of Voltaire*," Darwin begins, "& he insists strongly that direct attacks on Christianity (even when written with the wonderful force & vigour of Voltaire) produce little permanent effect: real good seems only to follow from slow & silent side attacks."[3] Johnston also quoted T. H. Huxley, Darwin's friend and "bulldog." "In addition to the truth of the doctrine of evolution," wrote Huxley, "indeed, one of its greatest merits in my eyes, is the fact that it occupies a position of complete and irreconcilable antagonism to that vigorous and consistent enemy of the highest intellectual, moral, and social life of mankind—the Catholic Church."[4]

My encounter with these two quotations gave substance to my original surmise that Darwinism is to no small degree a program designed by its creator for excising God, and was championed by a "bulldog" of the same mindset. Mind you, Huxley did not even share Darwin's belief that natural selection could explain all life-forms.[5] The essential thing for him was that Darwinism removed God from the picture. As Richard Dawkins put it a century later, "Darwin made it possible to be an intellectually fulfilled atheist."[6]

Several years after ordination, I was given the privilege of doing further graduate studies, this time in the history and philosophy of science. Although my primary interest was physics, there were breadth requirements in the program, so I enrolled in a seminar on the history of evolutionary biology. The seminar was lively, and the professor was enamored of Darwin. It was just a few years after the publication of Phillip Johnson's *Darwin on Trial*, so I used to take a copy of it to class. Whenever a topic of discussion arose in the class that Johnson had addressed in the book, I would read a snippet of it for the benefit of the others around the table. The usual response from the professor was to admit that Johnson had a good point and then to change the subject. By the end of the term, three out of the eight students thought that there was about as much truth to Darwinism as there was in Rudyard Kipling's *Just-So Stories* about how the leopard got his spots, the camel his hump, the elephant his trunk, and so on.

I have kept up an interest in the field of evolutionary biology ever since, and went on to get a PhD in the history and philosophy of science. I am now more convinced than ever that Darwin's theory can at best account for a small part of the variations in plant and animal forms throughout the ages. Its primary purpose is to serve as a creation myth for the secular society in which we live. In this role it contributes to an impoverished view of humanity and society, and is a constant threat to the Christian understanding of the creation and fall of man. Moreover, the thought patterns it engenders are destructive of common sense and of a responsible use of reason. It is, I contend, an unmitigated intellectual disaster, whose myriad shortcomings need to be dragged into broad daylight.

In my exploration of Darwinism, there were times when I felt the power of its darkness. If the story of Adam and Eve was not true, then how could the Church be a divinely founded and guided institution? I was happy to allow that the story in Genesis might involve some poetic license in the details. But what if, in fact, humans were not created in moral perfection, nor freely chose to abandon the good to embrace sin? What if instead the true story of man's origin was that we arose from a single-celled organism over eons of mindless evolution, with the inclination to violence and selfishness baked in through millions of generations of survival of the fittest? It seemed intuitively clear that natural selection should favor the lustful and violent. Darwin's theory was so much simpler—the effects of original sin part and parcel of evolutionary development. And yet the dogma of humankind's fall from sinlessness into original sin was too central, it seemed to me, to discard without changing the whole faith.

But if my allegiance was to truth, then what? I must seek the truth. Was I underestimating the potential of geological timescales to drastically transform life? Was it not the height of hubris to dismiss Darwinism when so many scientists and my peers had accepted it as true? Initially, I had little idea of molecular biology. Perhaps there was something hidden there that could account for apparent design apart from actual direct design. Could not God have created the various life-forms using the Darwinian mechanism? To be sure, God's ways are higher than man's ways, and no human should presume to know from unaided reason how God might choose to accomplish some act of creation. At the same time, even if it were possible, employing only the Darwinian mechanism to create the great variety of life seemed so inelegant and wasteful, and it left God remote and uncaring: an absentee watchmaker.

I resolved to give the theory a hearing. There were, after all, the obvious similarities between chimpanzees and humans. Perhaps there was no ontological leap between them and us, just a matter of degree, not a difference in kind. Moreover, many learned Catholics practically revered the theory. Through the 1960s and into the early '70s, Pierre Teilhard de Chardin was the rage. Even Joseph Ratzinger was cautiously appreciative of Teilhard and was looking for ways to accommodate Catholic theology to evolutionary thought. Was I holding on to an outmoded theology?

The best way to find the answers to my questions, I decided, was to explore the scientific evidence related to Darwinism. That was one of my motivations in signing up for the seminar in the history of evolutionary biology. This led me to keep reading material in the field long after the course was finished some thirty years ago.

I spent many years sorting through the evidence, aided by my graduate training in the history and philosophy of science, philosophy more generally, and logic, all of which I have found indispensable for navigating the various forms of evidence and arguments that impinge on the case for Darwinian evolution. Often I would encounter scientists deeply knowledgeable about the relevant biology but who seemed wholly innocent of the rules and methods of reasoning logically and avoiding fallacies. Some, for instance, would advance a wholly circular argument for this or that point of evolution and do so with no apparent awareness that they were committing one of the most elementary fallacies. At the same time I encountered individuals with a good grounding in philosophy and logic but who had not taken the time to master the foundational details of evolutionary theory. I saw that, given my background, I was well positioned to avoid both these shortcomings, provided I put in the spade work to learn more of the scientific debate, including more about the contemporary variations on Darwin's theory. And so weeks of study turned into months and years. The pages that follow are the fruit of that labor.

I have been planning to write this book for a long time, to help those who are searching for the truth about God and man in the face of Darwinian darkness. Phillip Johnson's *Darwin on Trial* (1991) was a great help to me, but it did not address some key questions of theology. Further, it is important to use up-to-date scientific evidence in the technical discussions, since much of relevance has been discovered in the intervening decades.

I also thought that it was important to argue for the perennial validity of intelligent design reasoning. To jettison this approach is to cut oneself off from biblical and patristic reasoning about God and creation. The hermeneutic of continuity, so often stressed by Pope Benedict in his reading of the Second Vatican Council, demands a defense of intelligent design. It is certainly legitimate to drop particular teachings of science, such as the model of an earth-centered universe, when they are shown to be untenable; it is another matter to drop the notion that the order and complexity of the universe demand a designing intelligence. To be clear, intelligent design as its chief proponents define and employ it does not get you all the way to theism—at least when the object of study is limited to biology. But its insights are most certainly theism-friendly, and if intelligent design is rejected, one is left cut off from a path to rational belief in God that was championed in the patristic tradition.

There are other questions that need to be addressed, because, along with the order and complexity of life and the universe, we also encounter suffering and death. Indeed, there is evidence that Darwin was spurred to develop his theory of evolution by his considering the presence of suffering and death in the world. He regarded this as sufficient reason alone to reject belief in a benevolent Creator. Having done so, he needed a substitute Creator, and he found it in the stuff of random variation and natural selection.

But in creating a new God, had Darwin also created a radically new understanding of man? It would seem so, and it is a big part of what makes the whole Darwinian controversy so passionate. Is the human soul a reality that only God can create? Or is it something that arises out of matter and the blind process of evolution? Is there something different in kind about us humans that distinguishes us from all other animals? Or are we, as Darwinian materialism holds, at bottom nothing more than meat robots?

Recently, a faithful catechist in the parish, who has been preparing children for their first Communion for many years, came to see me. "Father Martin," she said, "I have been teaching children about Adam and Eve, just as the *Catechism* tells us. But we can't be expected to believe that, can we? What is the real story?"

Her question was the immediate catalyst for me to sit down and write what I have learned on the subject over the years. Most of what will appear in the following pages has been said elsewhere by others, but I hope that putting it all together, at a level that educated non-specialists can grasp, will serve a useful purpose. It is time to dispel the darkness of Darwinian materialism.

When Hard Science Turns to Quicksand

Many years ago, when I was studying engineering, my friends and I would laugh at the "artsies." We *knew* that we were dealing with real knowledge whereas the artsies were forever stuck in the realm of contentious opinions. Every professor of electrical engineering could tell you whether a particular circuit would work; and every aeronautical engineer would come up with a similar answer about the lift of a particular airfoil. But try asking philosophers, theologians or, say, literary theorists basic questions in their fields, and there would be little consensus.

My friends and I used to dream about saving the university cartloads of money by getting rid of all the humanities departments. In this we were unreflectively aiding and abetting an idea known as scientism, the view that the hard sciences offer the only genuine knowledge of reality. Over the years, at least some of us have come to appreciate that the humanities produce disagreements precisely because they deal with the more fundamental questions of human life, which, unlike scientific theories, do not lend themselves easily to empirical testing or mathematical analysis.

But it remains true that the "hard" sciences tend to produce results which, in many cases, all their practitioners can agree on, and this surely redounds to the credit of these sciences as modes of pursuing knowledge, if we avoid the excesses of scientism. We can go further and affirm that biology is one such hard science. The discovery of DNA, the deciphering of the genetic code, DNA splicing, and the now routine sequencing of genomes of many different organisms are the result of the concerted efforts of many brilliant and dedicated people. All scientists who work in the field agree on these basic technical facts, which must ground, or at least be consonant with, biological observation of actual life-forms. And there is also much that "wet" biology—work done in the field—has discovered: the life cycles of the sand dollar and sea urchin, the chemical warfare of the bombardier beetle, the migrations of birds, salmon, eels, and turtles, the role of chemical identification in colonies of insects, etc. All these complex findings, once research teams have spent years studying them, can be understood and form the basis for broad agreement, even while inspiring awe of the living world around us.

There is, then, no doubt that the technical achievements of biology put it squarely into the league of hard sciences such as physics and chemistry.

But there is a field of biology where there is precious little agreement, and where even the limited "agreement" is far less about evidence than intellectual orthodoxy: origins biology and, specifically, evolutionary biology. Universities and scientific journals try to minimize the areas of uncertainty and dissent, but sometimes the truth comes out. As an example of a kowtowing to the Darwinian orthodoxy, consider the following statement from the website of BioLogos, a self-proclaimed Christian enterprise promoting a marriage of Darwinism and Christian theism. After defining evolution to be the theory that all present life-forms have descended from one original living cell via gradual changes over geological time scales, i.e., the Darwinian scenario, it goes on to state:

> There is very little debate in the scientific community about this broad characterization of evolution (anyone who claims otherwise is either uninformed or deliberately trying to mislead). The observational evidence explained by common ancestry is overwhelming. Of course new data causes scientists to adjust some of the specifics (like how long ago species diverged, or which species are most closely related), but this core view is overwhelmingly supported and agreed upon by the vast majority of scientists in the field.[7]

But then consider the following: Jerry Fodor and Massimo Piattelli-Palmarini describe themselves as "outright, card-carrying, signed-up, dyed-in-the-wool, no-holds-barred atheists." Their co-authored book is titled *What Darwin Got Wrong*, and it is clear that they mean something more than minor details. As they explain:

> This book is mostly a work of criticism; it is mostly about what we think is wrong with Darwinism. Near the end, we'll make some gestures towards where we believe a viable alternative might lie; but they will be pretty vague. In fact, we don't know very well how evolution works. Nor did Darwin, and nor (as far as we can tell) does anyone else. "Further research is required," as the saying goes. It may well be that centuries of further research are required.[8]

I have deliberately chosen counter-intuitive sources for these pro and con quotations. The more obvious option would have been to cite an atheist such as Richard Dawkins to champion the health and strength of the case for modern evolutionary theory, and the devoutly Catholic Michael Behe to underscore its shortcomings. But the fact that it is possible to find praise of Darwinism from Christians and far-reaching criticism from atheists underscores the difficulty facing anyone seeking clear answers regarding the status of evolutionary theory.

There are, of course, many books written on the theory of evolution, both pro and con. So one may legitimately wonder whether there is a need for yet another one. Has not everything that needs to be said on the subject been said? Although some arguments date back to Darwin's day, such as what to make of the lack of transitional fossils, each new scientific discovery reignites the debate between Darwinists and their critics. So there will always be a need for more up-to-date reports on the status of evolutionary theory. But my main concern is to help Christians understand what is at stake in the debate. My intention is to put into one volume diverse discussions that all have a bearing on assessing what science teaches us about the origin of species and especially about the origin of our species, *Homo sapiens*. At the same time, I intend to look at the implications of accepting Darwinism for our ability to know objective truth, which, of course, has a huge impact on philosophy. Finally, I want to look at the consequences that Darwinism has for Catholic theology.

To be clear, I never thought that the Bible was meant to be a science textbook. I am perfectly happy to accept that some fourteen billion years have elapsed since the Big Bang, and I find much evidence of deep time in physics, cosmology, and geology. I also am convinced that the testimony of nature, rightly interpreted, should be relevant to one's religious commitment. I could not, for example, imagine myself believing in monogenism—that we are all descended from one first couple—if there were conclusive reasons coming from science showing that our species never comprised fewer than ten thousand individuals. I take seriously the biblical notion that the order of nature is a powerful starting point towards coming to know nature's Creator. I find ill-informed all attempts, such as those of Stephen Jay Gould, to separate science and faith into non-overlapping magisteria: science dealing with truth, and faith with values. Only those who do not think that religion could be true can feel free to divorce it from reason.

I am aware that there are many well-catechized and fervent Catholics who seem to care little in what sense, if any, evolution is true. They know that the Church teaches that the first human couple rebelled against God and that original sin is an essential teaching of the Faith. But they are in no way disturbed by an otherwise completely naturalistic explanation for man's origin. They do not take the claims of evolution seriously enough to see that they can dissolve the very basis of their Christian commitment. These Catholics are joined by many evangelicals and Orthodox of a similar mind.

My hope is to waken such Christians from their slumber, and to give those Christians who care deeply about empirical evidence something to help them assess the status of the theory of evolution.

Evolution is a multi-faceted subject not confined to the science of biology. It quickly escapes its scientific confines and becomes a new metaphysics. It provides a new worldview persuasive to many. It should not come as a surprise, then, that it is going to be hard to unravel the empirical science from its wider cultural implications, interpretations, and influences. The threads do not easily come apart.

After the present introductory chapter, we will look, in Chapter 2, at the different possible meanings of the term *evolution*. Does it mean just change over time: transformation? Sometimes. More often it also encompasses an explanation of how the transformation comes about. We will need to look at the evidence for what kinds of transformations have actually occurred.

Chapter 3 will introduce an alternative account of biology: the theory of intelligent design (ID). This theory formalizes the steps by which investigators routinely detect design, in fields as diverse as forensics, archaeology, code-breaking, and data encryption. A seminal work in this field, *The Design Inference* by William Dembski (Cambridge University Press), offered a way to apply this reasoning in general. To appreciate the novelty of this approach to biology, we will look at a brief history of the philosophy of science, one of my areas of academic specialization. This will help us see how science has come to be presented as a materialist enterprise that regards any and all spiritual realities as vestiges of a pre-scientific mindset. This historical flyover will give us the tools to assess the legitimacy or illegitimacy of this shift in how science qua science is understood.

In Chapter 4 we will look at how ID—or, more broadly, teleological thinking—supports reason's path to knowledge of God's existence. A reflexive dismissal of the teleological argument undermines this path. It is an article of the Catholic faith that some such path from reason is possible, which is one of the seemingly paradoxical teachings of the Church: that we know by the certainty of faith that human reason, apart from revelation, can attain certainty about God's existence. The proponents of ID freely admit that their reasoning does not arrive directly at an omniscient and omnipotent God, but only at a mind who can act upon matter. The theists among us will of course make the next jump to God, motivated by other considerations, such as the moral law within us or a commitment to the possibility of metaphysical knowledge.

But a problem nevertheless remains, in that some biological systems appear to be the products of a sadistic mind: for example, the parasitic wasp who lays eggs in caterpillars so that her offspring can feed on live meat, or the mating antics of the black widow, who consumes her sexual partner. What does one say about all that? And what does one say about designs that appear to be clumsy or botched? These questions will be explored in Chapter 5.

In Chapter 6 we turn to the question of human evolution. What do we really know? What is bold speculation? The genetic similarities between humans and chimps are often cited as powerful evidence of common descent from an ape-like ancestor. Is this evidence really so clear-cut, or is there bluffing or fallacious reasoning involved? And what is the testimony of the fossil record? The question of human origins is, of course, a fundamental matter.

The evolutionary origin of man is taken to be the gospel truth in our secular society. Any attempts to criticize it bring about fast and furious retaliations. The usual story is that we came down from the trees, began walking upright, and then grunted our way to higher consciousness and the use of primitive tools and eventually language, each step driven by evolution's twin-mechanism of random mutations and natural selection. But what does such a scenario say about God's justice in condemning Adam and his descendants—and the visible creation with him—for some transgression that he could hardly even understand and dictated largely, or even exclusively, by a nature bequeathed to him by the shaping process of Darwinian evolution?

We need to look at the genetic record along with fossil evidence for early man. But it will also be necessary to look at the cultural achievements of our ancestors as evidenced in cave art, tools, and musical instruments.

Our next focus, in Chapter 7, will be the biblical conception of man as created in the image of God. Most moderns think that the mind is just an aspect of the brain. But there is considerable evidence in support of the contrary view, evidence that has poured in from diverse sources: Wilder Penfield's open brain surgery on conscious patients; the placebo effect; the voluminous documentation of near-death experiences, much of it inexplicable on materialist grounds and not easily dismissed as phantasms or fabrications; Thomistic epistemology; the nature of human language; and insights into human intelligence gleaned from advances in artificial intelligence. Any one of these lines of evidence should give materialists pause in their enthusiastic denial of the human soul as an immaterial reality.

In Chapter 8 we turn to the role that worldview played for the great proponents of naturalistic evolution. Darwin is usually portrayed as an inquisitive scientist, following truth wherever it led him. But is that accurate? Or was he an anti-theist in search of evidence for his metaphysical stance? Was the influential evolutionist Theodosius Dobzhansky an orthodox Christian in any recognizable sense of the word? If not, what was his worldview and how might it have shaped, or limited, the possibilities he was open to? And what about the biggest Catholic evolutionist of them all: Pierre Teilhard de Chardin? Did he have a place for original sin in his account? More recently, what are we to think of George Coyne's defense of Darwinism? He insisted that science and religion are totally separate, but in light of evolution, he thought that we needed to rethink the traditional Christian view of God's omniscience and omnipotence. Kenneth Miller, too, who has been decorated with all sorts of honors by Catholic institutions, supports Darwinism. Do these thinkers present a picture that Catholic theology can adopt? And how cogent is the evidence that they present for Darwinian evolution?

A lot of nonsense has been said about the Catholic Church's view of evolution. After John Paul II addressed the Pontifical Academy of Science in October 1996, the *New York Times* ran the headline "Pope Bolsters Church's Support for Scientific View of Evolution." It is clear that in the mind of the reporter, the "scientific view of evolution" meant modern Darwinism.[9] But did the Pope really endorse Darwinism? Or was that just wishful thinking by the "progressive" elements in the Church?

This book will cover a lot of ground and a lot of disciplines. So I will finish with a chapter that summarizes what has gone on before—a recapitulation. That way, if any chapter becomes too technical, readers can at least figure out how it was supposed to have fit into the general argument. My conclusion is that the scientific evidence for the possibility of a first couple, created *de novo*, is solid, and that Adam and Eve are an essential part of the Christian teaching. Bishop Spong is absolutely correct when he says that the traditional story of redemption does not make sense in the light of Darwin.[10] But I conclude that redemption continues to make sense because it is Darwin who was mistaken.

I wanted to keep this a relatively short book that tries to make sense of evolution as it pertains to some basic Catholic truths. As a consequence, I tried to stick to the points under discussion, rather than comment more fully on certain topics. For example, in discussing ID and proofs for the existence of God, I did not try to assess the relative roles of the intellect, the will, and grace in the process of man's becoming certain of the existence of God. As interesting as this topic can be, my only concern was to show that by rejecting ID, one blocks a very powerful intellectual path to God. I freely admit that there are other paths from reason and that one might need grace to overcome the darkening of the intellect by the will, but my present purpose is to focus on the path that ID offers.

No doubt, philosophically astute readers will find other topics where I tread more quickly than they might desire, relying on common sense to fill in this or that step in an argument. If the book disappoints them on this score, I hope that it will at least whet their appetites to further investigate the evidence of intelligent design and make them properly cautious of the grand claims of Darwinism. If so, they will be less likely to start developing alternate theologies of original sin, ones that tend to hollow out the foundations of a perennial moral law and hence of a cultural heritage of freedom, justice, and human rights.

One might legitimately ask whether anyone today can be so conversant with all the sciences as to make an informed criticism about Darwinism, especially if that person has never formally studied biology. We might just as easily ask whether a specialist in biology has the necessary tools to assess a theory that rests on so many different scientific disciplines, and that employs for support certain philosophical assumptions that extend beyond the empirical. Although I have done considerable research to bring myself up to speed on the technical aspects of biology relevant to this debate, it is my PhD in the history and philosophy of science, along with my training in metaphysics, that will prove every bit as indispensable. The scientific specialist tends to make the mistake of assuming that all branches of science answer to the particular methodology of his scientific specialization. Not so. Philosophers of science, who must of necessity study the methodologies of the various scientific disciplines, know better. Having done so, we know that it is more accurate to speak of scientific *methodologies*, in the plural, and to recognize that these methodologies have changed over time and are subject to ongoing revision.

Having recognized this, the historian and philosopher of science is better equipped than many a scientific specialist to regard the claims of methodological materialism with a properly jaundiced eye. Methodological materialism was imported into origins science, and there's an argument to be made that it has worn out its welcome, having begun to pinch and limit scientific inquiry, and that it is high time to show it the door. In any case, such a question is not best adjudicated by a scientific specialist with a limited view of the broader field. It is better adjudicated by those specializing in the question of what is the nature, or natures, of science and the sciences. That is precisely the subject of the history and philosophy of science.

More broadly, the question of our origins is fundamental to understanding who we are and, hence, the purpose of our lives. It cannot be relegated to some experts in a specialized field of science or the history and philosophy of science. It is a question for all of us, and the lines of evidence can be grasped and evaluated by any patient, thinking person. We will undertake such a project in these pages.

We are in the midst of a culture war on many fronts, with Darwinism serving as the creation myth of the atheists. The proponents of materialism are mistaken, not because they do not know science, but because their claims go beyond science and, in some cases, fly in the face of scientific evidence they are surely aware of but choose to ignore or explain away.

A strong basis for my confidence to judge where the science about human origins actually stands are articles in prestigious journals that are usually sympathetic to Darwinism. For example, in May 2021, a review article on human evolution was published in the leading journal *Science*. It concluded thus:

> Humans are storytellers: Theories of human evolution often resemble "anthropogenic narratives" that borrow the structure of a hero's journey to explain essential aspects such as the origins of erect posture, the freeing of the hands, or brain enlargement. Intriguingly, such narratives have not drastically changed since Darwin. We must be aware of confirmation biases and *ad hoc* interpretations by researchers aiming to confer on their new fossil the starring role within a preexisting narrative. Evolutionary scenarios are appealing because they provide plausible explanations based on current knowledge, but unless grounded in testable hypotheses, they are no more than "just-so stories."[11]

The American Museum of Natural History summed up the review:

> Most human origins stories are not compatible with known fossils… the number of species in the human family tree has exploded, but so has the level of dispute concerning early human evolution… However, many of these fossils show mosaic combinations of features that do not match expectations for ancient representatives of the modern ape and human lineages. As a consequence, there is no

scientific consensus on the evolutionary role played by these fossil apes.... Overall, the researchers found that most stories of human origins are not compatible with the fossils that we have today.[12]

As the lead researcher for the paper, Sergio Almécija, admitted, "When you look at the narrative for hominin origins, it's just a big mess—there's no consensus whatsoever."[13]

His point about lack of agreement was not meant to support intelligent design arguments. But the fact of the matter is that there is no consensus and the stories the public is being told are not compatible with the evidence. Moreover, this is not an isolated instance of mainstream science admitting a problem with evolution. The reader will find other examples throughout the book to make it clear that Darwinism is not a well-founded scientific theory.

I am encouraged by several recently published books on the subject of Darwinism from authors who have come to the same conclusions as I have. In *Taking Leave of Darwin*, Neil Thomas gives his reasons for abandoning his belief in the materialistic paradigm he had so confidently accepted in his youth. A longtime member of the British Rationalist Association, he is not ready to adopt a religious faith. Yet he describes himself as a truth-seeker, and it is clear to him that Darwinism does not follow from the evidence:

> It appears that the ideological necessity of finding a strictly materialist theory eventually came to trump those honest and open-minded objections voiced by the majority of the reviewers of the *Origin* [*of Species*] in the decade after its publication. Intellectual integrity was sacrificed on the altar of ideological commitment. Natural selection was thrust forward aggressively like a form of profane crucifix to ward off the danger thought to be posed by religion, in a way functionally not dissimilar to Voltaire's oft-repeated rallying cry of 'écrasez l'infâme' (= "crush the infamous one," by which Voltaire meant the superstitions of religion). Darwinism was beginning to assume for some the function of an anti-religious apotropaic or magic charm against supposed evil.
>
> But as we will see... for an increasing number of investigators, the charm is wearing off.[14]

Another noteworthy contribution comes to us from Robert F. Shedinger: *The Mystery of Evolutionary Mechanisms*. Shedinger is a scripture scholar, but not one likely to inspire confidence among traditionally minded Christians. As he confesses, "My earlier books have the titles *Was Jesus a Muslim?* and *Jesus and Jihad*, so my skepticism of Darwinism cannot be explained by appeal to conservative Christian ideology!"[15] Perhaps I am partial to Shedinger's work on evolution because, like me, he did his undergraduate studies in science and engineering.

Shedinger took a year's sabbatical to prepare himself to teach a course in religion and science. He began his research with the assumption that "Charles Darwin had essentially solved the problem of the origin of species with his concept of natural selection, and that modern evolutionary theory was simply a more complex extension of Darwin's basic insight and stood as one of the most successful and empirically verified scientific theories ever proposed."

But after he read most of the landmark scientific papers and books in the field, he changed his mind. He asks, "If evolutionary theory is as empirically confirmed as most biologists say it is, and if ID can be easily disparaged and dismissed, why do biologists feel the need to create a caricature of the ID movement rather than simply facing it head-on and showing why it is wrong?" As he read the relevant scientific papers, he noted, "I was shocked when I began to recognize just how ambiguous and tentative so much of the literature is. It is littered with caveats, inconsistencies, unsupported assumptions, grand claims backed by a dearth of empirical evidence, and perhaps most surprising of all, by 'religious' terms such as 'orthodoxy,' 'heresy,' 'dogma,' 'creed,' and 'blasphemy.'"

The Mystery of Evolutionary Mechanisms contains an excellent history of evolutionary biology, and it illustrates exactly what Shedinger means by inconsistencies and unsupported assumptions. Ernst Mayr, one of the leading figures of evolutionary studies in the twentieth century, could accuse dissenters from Darwin of such colossal ignorance that refuting them would be a waste of time. Yet at the same time, Mayr could admit that the basic theory of evolution is "hardly more than a postulate and its application raises numerous questions in almost every concrete case."[16] Lest one think that the situation has changed much, Shedinger cites the eminent biologist James Shapiro who, in the late 1990s, admitted that "evolution remains a mystery. Its fundamental driving forces have not been resolved either in detail or in principle."[17]

Several books have also appeared that deal with the more circumscribed topic of human evolution, which recognize that there are conflicts between the standard Darwinian account and the traditional Christian doctrine. *At the Dawn of Humanity: The First Humans* is by the Catholic biologist and philosopher of science Gerard M. Verschuuren. He accepts the evolutionary scenario for the human body but respects the Church's teaching on monogenism and original sin. I am pleased to note that he uses many of the same arguments for the spiritual nature of the soul that I present in this book.

William Lane Craig is a Protestant philosopher who recently published *In Quest of the Original Adam*. While his convictions on the question of evolutionary theory are unclear, in the book he accepts for the sake of argument that the Darwinian evolutionary account is true in the main and seeks to show Christian readers how traditional Christian teaching can nevertheless be safeguarded in light of that assumption. He spends much time trying to discern what the Bible has to say. He sees the importance of an original pair and so posits an original pair; but he does not say that Adam and Eve are our *sole* progenitors. In order to make sense of some genetic data and to harmonize his account with the evolutionary theory, he is willing to allow the possibility of their breeding with some human-like apes.

I am pleased to see that there are efforts by these serious scholars to argue for the traditional Christian belief in Adam and Eve. But ultimately, I think that they are going about it the wrong way. Darwinism is an acid that dissolves all common sense, that destroys all confidence in our ability to attain truth. It is not enough to erect a makeshift dam around Adam and Eve. A much better strategy is to slay the hydra once and for all, not just because it is a danger to a Christian dogma, but because it is an affront to human reason, without which neither science nor Christian theology is possible.

It is my contention that common sense plays a strong role in adjudicating the evidence for Darwinism, so the non-technical reader need not feel intimidated. There are sciences where common sense might fail in some instances. Time dilation in special relativity or the behavior of electrons in a double-slit experiment in quantum mechanics leap to mind. But so far, I have found nothing in biology to shake my belief that the apparent design in nature is real design and that there is a Designer to account for it. And the good news is that this Designer is calling us to share in his intimate life, although our first parents rebelled against Him.

2. Evolution: More than a Hypothesis

> *Today, nearly half a century after the publication of the encyclical [Humani Generis (1950)], new knowledge leads to the recognition of the theory of evolution as more than a hypothesis.*[1]
> —John Paul II, Address to the Plenary Assembly of the Pontifical Academy of Sciences (1996)

> *I would like to inform the Holy Father, however, that he has been misinformed.*
> —Giuseppe Sermonti, Chief Editor of *Rivista di Biologia* / Biology Forum[2]

What is the origin of all the plants and animals we encounter in the world around us? One answer comes from a common reading of the book of Genesis: Each plant and animal today is just a more recent copy of its first ancestor, which was directly created by God. In this scenario, there is no evolution, at least none to produce fundamentally new *kinds* of animal forms; and that was the consensus until the eighteenth century. To be sure, there was some speculation in the Judeo-Christian tradition about whether and to what degree animal species could change over time. In late antiquity, Augustine famously introduced the notion of "seminal reasons" to explain how God could be said to create at the beginning things that would only be visibly present in later ages. Just as seeds develop into plants over time, so one created reality could give rise to another as history unfolds.[3] In the fifteenth century, the rediscovery of Lucretius's long poem *De Rerum Natura* (*On the Nature of Things*), presenting an atomistic/materialistic/evolutionary account of the origin of life, may have sown the seeds of materialist evolutionary speculation in Europe, though if so, the seeds were very slow to sprout. The possibilities of plants and animals changing radically over time did not capture the imagination of the learned in Christendom until the middle of the eighteenth century.

Modern speculation about the changes in life-forms over the ages can be traced back to the book *Telliamed*, by Benoît de Maillet (1656–1738). The book, whose title is "de Maillet" spelled backwards, was published posthumously in 1748. Today one hardly ever hears of de Maillet, but *Telliamed* was influential in the second half of the eighteenth century. It was disparaged by theologians and some men of science. Nevertheless, as the *Encyclopedia of Philosophy* notes, "Comte de Buffon, Denis Diderot, Chevalier de Lamarck, and Erasmus Darwin, among others, availed themselves of Maillet's theories as a starting point for even more daring concepts of their own."[4]

Georges-Louis Leclerc, Comte de Buffon (1707–88) speculated that species may have improved or degraded as they moved away from their centers of creation. Erasmus Darwin (1731–1802), the grandfather of Charles Darwin, published *Zoonomia* (1794), in which he proposed "that all warm-blooded animals have arisen from one living filament, which THE GREAT FIRST CAUSE endued with animality." He did not, however, provide any explanation as to how this might have come about.

Jean-Baptiste Lamarck (1744–1829) suggested that change could happen through the inheritance of acquired traits. In this scenario successive generations of, for instance, an antelope-like creature would be born with longer and longer necks as their line of progenitors strove to find their food in increasingly higher branches, each generation stretching its necks ever higher, eventually giving rise to a new species, the giraffe.

Charles Darwin conceived his theory of evolution in the late 1830s but only went public with it in 1858, after he had received a letter from Alfred Russel Wallace which described much the same ideas about evolution that Darwin had been secretly coddling for some two decades. Darwin's friends Charles Lyell and Joseph Dalton Hooker convinced him to publish his theory; and the two of them wrote a letter to the president of the Linnaean Society about the theory, which, they explained, was first conceived by Darwin but also independently arrived at by Wallace.[5] Darwin published his *Origin of Species* in 1859; and it is his name that we now associate with evolution. Darwin was the first to conceive of the theory; and, unlike Wallace, he came from the upper echelons of British society. His mother was from the immensely wealthy Wedgewood family; and he married his cousin Emma Wedgewood. Wallace was abroad doing field work whereas Darwin was right in the heart of England, and as a gentleman of independent means had considerable leisure time to advocate for his theory in subsequent books and through a voluminous correspondence with top scientists in England, Europe, and the United States. Finally, Darwin insisted that his theory could account for all life, even for all the faculties of human beings, a view Wallace ultimately rejected. For all these reasons Darwin, rather than Wallace, became the clear-cut choice for patron saint of modern evolutionary materialism.

Today when the subject is discussed, it usually flies under the simple term *evolution*. But what exactly is meant by the term? It may refer to nothing more than the belief that today's flora and fauna descended from different flora and fauna in the past. In the nineteenth century, canal building in England unearthed many different fossils, some clearly of animals and plants that were long extinct. People soon noticed that there was some order to the fossils: the same succession of layers was found in different parts of the country. And the different characteristic layers tended to have the same kinds of fossils. Assuming that deeper layers were more ancient and the more recent layers were closer to the surface, it became possible to reconstruct a relative timeline of the flora and fauna over fairly wide geographic regions. One hypothesis that immediately came to mind was that earlier flora and fauna gave rise to more recent flora and fauna. If there is no further speculation as to how this happens, then we are left with one meaning of evolution, which we might call transformism. Very few people find this meaning of evolution controversial, especially if the transformation stays within certain boundaries.

I hold a longer discussion of the term *transformism* for Appendix A, because I do not want to get sidetracked here by discussions of whether transformism entails universal common descent, how quickly transformation happens, and possible causes. Transformism serves at this point to describe the well-founded hypothesis that at least some of today's present life-forms arose from past life-forms.

The fossil knowledge from the nineteenth century as well as some more recent genomic data are probably what Saint John Paul II had in mind when in 1996 he stated that "some new findings lead us toward the recognition of evolution as more than a hypothesis." The popular press summarized John Paul II's address to the Pontifical Academy of Science as papal acceptance of Darwinism, but such a conclusion could be arrived at only by those who cannot conceive that there might be any difference between evolution and Darwinism. It also betrays their inability to read—or to be honest—because, in the very same short address, the Pope said: "And to tell the truth, rather than speaking about the theory of evolution, it is more accurate to speak of the theories of evolution. The use of the plural is required here—in part because of the diversity of explanations regarding the mechanism of evolution, and in part because of the diversity of philosophies involved."

Clearly the Pope was persuaded that evolution, in the sense of change in life-forms over time, had occurred. This is the first and more generic meaning of evolution, what I referred to as transformism. At the same time, the Holy Father was aware that there is controversy about the process of this change. It is a great triumph of materialist propaganda to equate the word "evolution" in the public mind with a particular explanation for how this change happened. Darwin became famous precisely because he proposed a purely natural and mindless process which, he believed, could account for the huge diversity of flora and fauna over the whole history of life on earth. The reader interested in further exploring the various meanings of evolution would do well to consult Jay Richards's introduction to *God and Evolution* or Stephen Meyer's and Michael Keas's "The Meanings of Evolution."[6]

In the *Origin*, Darwin began by pointing out that in nature there are variations among individuals of a species. Everyone agrees with this, because these variations are what enabled humans to develop various breeds of farm animals, dogs, pigeons, fruit trees, vines, etc. Breeders notice individual plants and animals exhibiting some advantageous traits and then selectively breed these to develop new strains. Humankind has been quite proficient in this enterprise.

Since Darwin was looking for a materialistic explanation for biological evolution, he needed to replace the intelligent breeder with a natural means of selecting these random variations. To do this, he could point to two more uncontroversial facts. There are limited resources in the world; and every living plant and animal produces many more seeds and offspring than can possibly survive and reproduce. A female salmon, for example, lays a few thousand eggs. On average, only two of these will become fish that will come back to reproduce in the same stream bed several years in the future.

Darwin reasoned that the innumerable life-forms are locked in a struggle to get at the finite amount of resources. And this life-and-death struggle selects any advantageous trait that might spontaneously appear. Yes, the human breeder is intelligent. But natural selection never goes to sleep—it is ever vigilant—and it has been around for a long time: thousands, millions, even billions of years.

The idea is inarguably clever, and whatever becomes of Darwin's grander claims, the attention he brought to the role of natural selection in biology is an undoubted contribution to science. At the same time, any theory of evolution that depends significantly on natural selection to seize upon and lock in a series of small evolutionary variations in a population comes at a price: each small step must be functional and, in most cases, some sort of improvement in fitness. Yoking a theory of evolution to natural selection met with skepticism from friend and foe alike. Even for a Darwin champion such as Huxley, this price was too high, since he saw little evidence that nature could afford such a convenient series of small steps from one form to a fundamentally different one. So skeptical was Huxley that that he urged evolutionists to disentangle a general theory of evolution from a close dependency on natural selection.[7] (Such skepticism has resurfaced in the modern period with the evolutionary school that emphasizes the role of what is known as *neutral evolution*.[8])

The reception of Darwin's theory, as interesting as it is, need not detain us here. But it is important to mention one other serious early challenge before moving on. Unknown to Darwin (and to most everyone else at the time), the Augustinian priest Gregor Mendel published his experiments on the breeding of peas in 1866 in an obscure journal. His work was discovered in 1900 by scientists who had independently made the same kind of observations, which were unfavorable to Darwinism. The challenge to Darwin's theory came through the discrete nature of the genes that Mendel had discovered. In breeding peas, he focused on seven traits. Each of these seven traits came in two forms. The color of the seed, for example, was either yellow or green; the color of the flower was either purple or white. There was no in-between, i.e., no blended inheritance. And it was clear that the genes (or "factors," as Mendel called them) persisted even though the parent plants did not have the trait. So, for example, he found that a purple flower, when self-pollinated or even when pollinated from another purple flower, can produce white flowers, but there was no pink. This is a case of discontinuous inheritance.

Darwin, on the other hand, imagined that every part of the parents' bodies produced "gemmules," or little cells, which the blood stream would transport to the eggs and sperm; and these gemmules would then go on to form the body of the offspring. It was clear that Darwin's genetic theory could not explain how two purple flowers could give rise to a white flower or why a purple and a white flower did not produce a pink flower. So, at the beginning of the twentieth century, Mendelian genetics posed a significant problem for Darwin's theory of evolution.

There is a saying that "with statistics you can prove anything"; and it was from statistics that salvation came for Darwin. Mathematically minded biologists shifted their gaze from the individual to the population. In 1930, Ronald Fisher published *The Genetical Theory of Natural Selection*, a work of advanced statistics. Most working biologists could not begin to understand it, but they took his word that he had done for biology what Isaac Newton had done for gravity: tied everything neatly together. Yes, genes persist, as Mendel had shown, but once in a while a chance mutation in a gene occurs, and once in a very long time, the mutation is helpful. Thanks to natural selection these mutations stand a good chance of being passed into the population, and over the years an accumulation of such mutations changes the average individual of the species.

There was no biology in Fisher's book—no one knew about the molecules of heredity in 1930—but mathematics had spoken; and there is no arguing with mathematical rigor, particularly if you know very little mathematics and the underlying molecular biology remains a mystery. Soon two other mathematical geneticists, John Haldane and Sewall Wright, published works along the same lines, albeit with some major technical differences.[9]

According to Ernst Mayr, the great unification of evolutionary thought came through Theodosius Dobzhansky's *Genetics and the Origin of Species* (1937). Unlike the mathematicians, Dobzhansky had done much biological work in laboratories, so he could unite the insights of the naturalists and the population geneticists. By 1947 the diverse branches of biology "came together and declared their acceptance of this newly synthesized, integrated evolutionary theory."[10]

This "Modern Synthesis"—neo-Darwinism as it is sometimes called—did not convince everyone, but it had a very powerful propagandist, Julian Huxley, the grandson of Darwin's mouthpiece and bulldog Thomas Huxley. Speaking at the 1959 celebrations of the centenary of *The Origin of Species*, Julian Huxley said:

> In the evolutionary pattern of thought there is no longer either need or room for the supernatural. The earth was not created, it evolved. So did all the animals and plants that inhabit it, including our human selves, mind and soul as well as brain and body. So did religion.... Finally, the evolutionary vision is enabling us to discern, however incompletely, the lineaments of a new religion that we can be sure will arise to serve the needs of the coming era.[11]

Huxley was the 1950s version of Richard Dawkins. Neither Huxley nor Dawkins leaves any room for God. The difference is that Huxley did not consider religion a pejorative word.

The Grand Theory in Light of the Empirical Evidence

Darwinism (I will dispense with the "neo" and henceforth allow context to indicate whether I am referring to Darwin's original theory or the updated version summarized immediately above) is an all-encompassing answer to the question of biological origins. But is it true? Is there empirical evidence for the theory? I have come to the conclusion that Darwinism as a grand theory cannot be defended on scientific grounds. And I hope to make that obvious to you.

So how did I come to this conclusion? As mentioned earlier, I took a graduate course in the history of evolutionary biology in the early 1990s. It was just a few years after the publication of Phillip Johnson's *Darwin on Trial*, so I thought that it might be interesting to review the book as one of my class assignments. Not being a specialist in biology, I read reviews of the book by those more learned in the field. Harvard paleontologist Stephen Jay Gould, for example, criticized Johnson's writing style and then went on to pontificate about there not being even the possibility of a clash between religion and science (in this case, evolution) because science was about "factual reality, while religion struggles with human morality."[12] David Hull went on a rant that he wanted nothing to do with a God who creates insects who poison, decapitate, and devour one another, so Darwin must be right and Johnson wrong.[13] Although both Gould and Hull brought in some arguments from science, it was clear that both were livid because Johnson had effectively attacked their philosophical worldviews. It was also clear to me that neither had as strong a scientific rebuttal in their back pockets as they assured readers; otherwise they would have felt no need to resort to philosophical and theological arguments, and dodgy ones at that. Their responses were enough to convince me that Johnson might be on to something after all. Perhaps the scientific case for Darwinism was indeed shockingly weak. This sense has only been reinforced by my many subsequent years of study.

Darwin was influenced by his friend, the well-known geologist Charles Lyell. Darwin took Lyell's *Principles of Geology* on his voyage on the Beagle and imbibed its uniformitarianism: a doctrine which states that the present is the key to the past. In geology, we need not posit some ancient catastrophe to explain how mountains were made.

Mount Etna, for example, can be explained by the regular eruptions of lava which have built up the volcano a few hundred feet at a time over spans of hundreds of thousands of years. One might think that a volcanic eruption is fairly spectacular, even catastrophic; nevertheless it is a process that we see now and which we can use to explain the events of the distant past. So it provides a uniformitarian explanation.

In biology, uniformitarianism demands that nature make no jumps. Darwin called it a "canon" of natural history. And the phrase *natura non facit saltum* occurs eight times in *The Origin of Species*. Why this insistence? We have no experience of a great jump in biology since the dawn of history, so any significant jump would immediately be attributed to a miracle, and not to a law of nature. So Darwin, true to his uniformitarian principles, always insisted that the variations that nature selects must be small.

If Darwinism is true, then we should see only smooth transitions among the various plant and animal forms. That is obviously not the case at the present time. True, there are many instances of species so similar one to another that they spark debates among the lumpers and splitters as to whether two types are distinct species or just breeds of the same species. Yet today's flora and fauna tend to be quite easy to slot into distinct categories, if not at the species level then certainly at the level of genus, family, order, and so on. Here the gulf between distinct groups is unmistakably large. But how about life in the past?

The Fossil Evidence

The tree of life is the only picture in the *Origin*, and is a well-known icon of evolutionary theory. But we will use Ernst Haeckel's version because it makes clearer some of the notions that Darwin wanted to illustrate. The vertical axis, the height, represents the time elapsed since the serendipitous arrival of the first self-replicating life-form in some warm pond eons ago. This single-celled life-form is the base of the tree. As time progresses, we move up the page, seeing what looks like a trunk. Then we start to see branching when two population groups get separated and each adapts to the local conditions. Each of these then gets split up, causing more branching. As we go higher up the tree and do a horizontal cut, we get progressively more species, with the differences between them being proportional to the horizontal distance along the cut. But is there evidence for this tree?

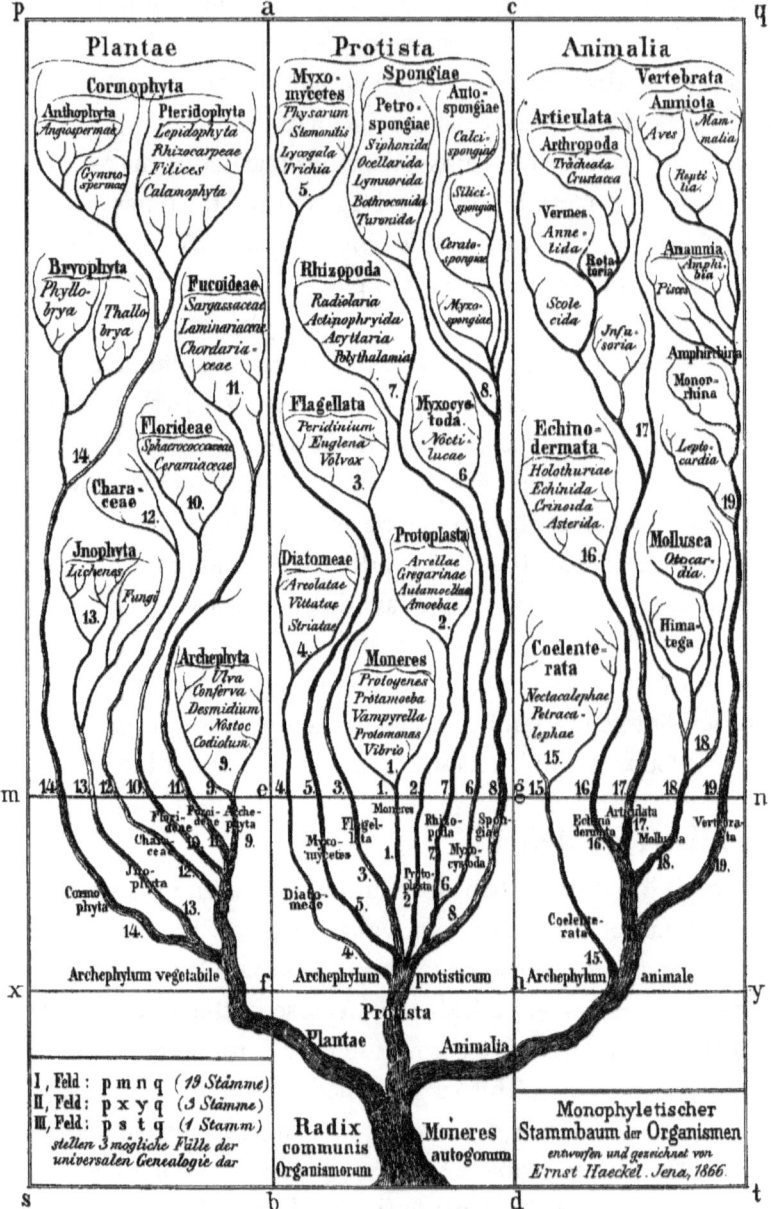

Figure 2.1.
The tree of life as drawn by Darwin's German defender, Ernst Haeckel.

The obvious way to check to see what happened in the past is to examine the fossil record. If Darwin was right, we should be able to find fossils leading from one species to another by almost imperceptibly small steps, so that if we took a picture of each of them and then quickly projected these pictures one after another, we should get a video showing the transformation, in the same way that time-lapse photography can show us a budding plant transform into a blooming flower in a manner of seconds. Darwin knew that the fossils gave little support to his theory, but he could console himself by appeals to the incompleteness of the fossil record. He hoped that in the future these gaps would be filled by new fossil discoveries. But these hopes have not been realized; on the contrary, the evidence against a slow gradual transformation of species has grown.

What about the series of horses, from a fox-sized animal to the modern horse? Do these not show smooth transitions? Perhaps there is a progression. But there are two points to keep in mind. First, does the progression prove descent? Imagine lining up the skeletons of the least weasel, mink, ferret, marten, otter, badger, and wolverine: all members of the *Mustelidae* family. Would that prove that the wolverine descended from the least weasel? In the case of the horse fossils, the fossils get bigger as they get more recent, so the evidence of descent is better. But if Darwinism were true, one would expect many more such sequences. And the fact that we keep seeing the horse sequence recycled time and time again tells us that such sequences are rare. The second point to keep in mind is that this sequence may indeed be an argument for Darwinian evolution, but it is fairly limited. In the case of the *Mustelidae*, Darwinian evolution, if real, does not transcend the level of family. Something much more is needed for the grand claims of Darwinism.

One body of fossil evidence that is particularly telling against Darwin comes from the fossils in and before the Cambrian era, and especially from a very narrow five million years within the Cambrian era, from 530 to 525 million years ago (Mya).[14] Darwin was aware that this era gave rise to a great variety of fossils for which there were no antecedents in earlier layers. As time went on, much richer sources of

Cambrian fossils were discovered. The first major find was the Burgess shale deposits in British Columbia, discovered in 1909. Since then another rich source was discovered in Chengjiang, China, in 1984. In a very short geological span, some 0.1 percent of the time since the earth was formed 4.5 billion years ago, a menagerie of life-forms arrived on the scene, without warning. This is known as the Cambrian explosion.

It is hard to say anything about this explosion without running into some opposition. The Darwinists instinctively know that it is powerful evidence against them, so they have circled the wagons. In *Darwin's Doubt*, Stephen Meyer describes the antics surrounding the showing of a documentary film, *Darwin's Dilemma*, at the University of Oklahoma. He and Jonathan Wells, from Discovery Institute, were the presenters. Since the public funding policies of the University did not permit it to deny a forum to groups such as Discovery Institute, the administration had to go into damage control mode. They issued a disclaimer saying that the university did "not support unscientific views masquerading as science." A lecture was scheduled by one of the professors to mock the Cambrian "explosion," which ended just before the film was to start. The question-and-answer period after the film was packed with a hostile crowd of experts insisting the Cambrian fossils posed no problem because their precursors had been found.[15]

It is true that there are some fossils from before the main event of 530–525 Mya, but they are far from sufficient to explain the astounding variety of forms that emerged during the Cambrian. A second tack is to argue that the Cambrian precursors must have been soft-bodied creatures, and soft-bodied creatures don't fossilize. But that is clearly not the case, because many of the Cambrian fossils were themselves of soft animals. University of Cambridge paleontologist and evolutionary biologist Simon Conway Morris thinks that to have escaped detection via trace fossils, any such putative ancestors would have had to be less than a millimeter in length, and what traces there are from this period may not even be from an animal, but rather from, say, "strolling protistan 'slugs,' analogous to slime-mold Dictyostelium."[16]

An important point to be clear about: The Cambrian wasn't merely an explosion of new species. It was an explosion of entirely new phyla or body plans. Phyla are a high-level classification in the Linnaean system. The categories, in order from most general to most specific, are domain, kingdom, phylum, class, order, family, genus, and species. According to Darwinian theory, it takes the evolutionary process much, much longer to create entirely new phyla than it does new species, and yet in the Cambrian more than twenty new phyla seem to appear out of nowhere. There are presently a total of thirty-six phyla in the Linnaean system; three arose in the Precambrian period; twenty or more, in the Cambrian; four, in later geological periods; and nine are represented only by existing forms.[17]

A problem with presenting this evidence is that it runs directly into a debate about classification schemes. Darwin believed there was a natural classification of all life that formed a record of evolution: a family tree. Be that as it may, any actual classification scheme can be challenged as being a human artifact. If, for example, a really strange creature is found that is too different from any other life-form, does that mean that all of a sudden one has to devise a phylum, class, order, family, genus, and species to give it a name? And, if so, what level of reality do any of these categories possess? The classifiers have their own biases. Some are lumpers and some are splitters. These are not going to agree on the number of phyla that arose during the Cambrian era.

To add to the confusion, there are some very strong advocates for a different kind of classification system, usually called cladistics (from the Greek word *klados* meaning "branch"). This system puts life-forms sharing certain derived characteristics together and tries to determine the evolutionary branching that took place since the last presumed common ancestor. There is no attempt to rank the clades as there is in the Linnaean taxonomy. Fortunately, we need not get dragged into this argument, for, as Meyer summarizes:

> The Cambrian explosion presents a puzzle for evolutionary biologists, not just because of the number of phyla that arise, but rather because of the number of unique animal forms and structures that arise (as measured, perhaps, by the number of phyla)—however biologists decided to classify them. Thus, whether scientists decide to use newer rank-free classification schemes [cladistics] or

older, more conventional, Linnaean categories, the "evolutionary novelties"—that is, the new anatomical structures and modes of organization—that arise suddenly with the Cambrian animals remain as facts of the fossil record, requiring explanation.[18]

Fodor and Piattelli-Palmarini concur with this assessment. In their criticism of Darwinism, they speak of morphological explosions and conclude, "We can summarize by saying that morphological explosions may well reflect major changes in internal constraints as crucial components in speciation. If so, then the effects of natural selection may well consist largely of post-hoc fine-tuning in the distribution of subspecies and variants: quite a different kind of account from the one of gradual selection of randomly differing small variations."[19]

Ironically, even the museum at the University of Oklahoma, which was so opposed to the talk by Meyer and Wells, had a display of fossils that illustrated the Cambrian challenge to Darwinism. As Wells describes it, "It showed over a dozen of the Cambrian phyla at the top of a branching tree with a single trunk, but none of the branch points corresponded to a real living thing."[20]

Given the controversy about the Cambrian explosion, one might think that it is the only major challenge to Darwin's theory in the fossil record. But it turns out to be one of many sudden bursts of new lifeforms. There is, for example, the Great Ordovician Biodiversification Event (GOBE), in which about 300 new families of marine invertebrates appeared between 485 and 460 Mya. There is the Odontode Explosion, in which all major groups of jawed fish with teeth or toothlike structures arrived on the scene between 425 and 415 Mya. The Carboniferous Insect Explosion occurred between 318 and 300 Mya. The different subgroups of amniote tetrapods arrived on the scene between 251 and 240 Mya. These include the dinosaurs, turtles, lizard relatives, croc relatives, and the first mammal-like animals. Flowering plants appeared and rapidly diversified between 130 and 115 Mya. Modern placental mammal orders appeared between 62 and 49 Mya, without known precursors. Modern bird groups arrived between 65 and 55 Mya. And there are several other explosions that Meyer and Günter Bechly describe.[21]

Paleontologist Stephen Jay Gould, no friend of Christianity or intelligent design, summarized the awkward situation for him and his fellow evolutionists:

> The extreme rarity of transitional forms in the fossil record persists as the trade secret of paleontology. The evolutionary trees that adorn our textbooks have data only at the tips and nodes of their branches; the rest is inference, however reasonable, not the evidence of fossils.... We fancy ourselves as the only true students of life's history, yet to preserve our favored account of evolution by natural selection we view our data as so bad that we never see the very process we profess to study.[22]

The available fossils, in fact, provide some arresting evidence against Darwinism, beside discontinuity. I think that most of us assume that mimetics are the product of natural selection. We are told that viceroy and monarch butterflies are mimics of each other, each trying to look like the other because each is unpalatable to some predator. It makes good Darwinian sense. Who would think to challenge it? I am not about to get into the debate about these butterflies as to which is mimicking which. Rather, I will share my surprise about another case of what is often passed off as selection-enhanced mimicry: stick insects, which belong to the order of phasmids. Surely these adapted themselves to blend into their background in ecosystems with lots of sticks and slender branches, right? It turns out there is a major problem with that idea. Giuseppe Sermonti explains:

> The oldest phasmid fossils (they go back in Baltic amber to the Tertiary—i.e., about 50 million years ago) look identical to present day species, showing that no gradations have occurred. It is thought that those phasmids originated from Chresmodids of the Upper Jurassic in Germany, fossils of which are encountered in deposits dating back some 150 million years. But the oldest fossils of stick or leaf insects (protophasmids) go back to even remoter periods, in the Permian (250 million years ago, in the Paleozoic). One might argue that these insects completed the process of imitating leaves at an extremely gradual rate beginning at a still earlier time. Yet things do not work out this way. Plants with flowers and leaves...

appeared no earlier than the Cretaceous—in other words about 100 million years ago, long after the first protophasmids. This chronological anomaly places the imitators earlier in time than the objects of the imitation, leaving entomologists and paleontologists disconcerted.[23]

There is one more technical point to keep in mind before we take leave of the fossil evidence. Darwin's theory requires a bottom-up tree. In his scenario, by the time that we get to something that taxonomists are willing to call a phylum, there should be numerous species, genera, and families all hinting at its eventual arrival. On the Darwinian view, species develop into genera; genera, into families; families, into orders, etc. Instead, the fossil record shows a top-down approach. A small number of life-forms suddenly arrive representing an entirely new body plan, a new phylum. Then, as time goes by, the phylum gets filled in with more and more representative species.

The history of life can perhaps be likened to a collection of different musical themes. In Beethoven's Fifth Symphony, the four beginning notes are the theme. They are then developed into many variations that still recognizably reflect the same theme. But there is no smooth way to get from the theme in his Fifth Symphony to the theme of the "Ode to Joy" movement of his Ninth Symphony. The fossil record, in other words, fits more the pattern of an artist creating a new theme and then working variations on it. Or to return to Darwin's proposed picture of an evolutionary tree of life, with all of life's variations branching out from a single ancestor "trunk," it seems that the history of life instead is more like a separate collection of bushes than a single tree.

Breeding Experiments

Darwin used the well-known achievements of breeders to argue that nature could replicate these successes and, given enough time, go further. So it made sense that biologists should try to confirm this experimentally. What have been the results of those efforts? In the more than a century and half since *The Origin of Species* appeared in print, the picture is little changed in this regard. There are a great variety of dogs and pigeons and sheep that have been bred by humans over millennia, but they are still dogs, pigeons, and sheep. All of these, however, reproduce slowly and are relatively large. So the experimentalists turned to a much smaller and faster-breeding animal: *Drosophila melanogaster*, the common fruit fly. The advantages are obvious. Thousands of specimens could be kept in jars in a single room. The generation time from egg to egg-laying female is around two weeks. And no one minds killing vast numbers of the offspring while selecting some interesting mutants to breed further.

The results in the lab were disappointing to the Darwinians. No new species were created. Sterility or reversion to type was the inevitable outcome. According to John A. Davison, Theodosius Dobzhansky "proved experimentally that artificial selection was unable to transform the fruit fly, *Drosophila melanogaster*, into a new species." Davison continues: "The reality is that *Drosophila melanogaster* can't change and, as far as we know, hasn't changed for millions of years."[24]

Hopeful Monsters

These negative results were known since the 1930s. In 1940, Richard Goldschmidt published *The Material Basis of Evolution*, in which he said that microevolution could not explain the major changes. So he put his hope in macromutations or saltational events, sudden, big, multi-mutational leaps forward. His detractors spoke of "hopeful monsters" and ridiculed and vilified him for decades afterwards. Stephen Jay Gould remembers his graduate school days at Columbia University in the early 1960s:

> I had never heard of Richard Goldschmidt. Yet his name surfaced in almost every course—never with any explanation of his views, but only in a fleeting derisive reference to something called a "hopeful monster." Students then responded with a derisive sign of recognition—as our professors seemed to expect as a badge of membership in some inner circle. I found the oft-repeated exercise—one might almost have called it a ritual—offensive and demeaning, both to Goldschmidt and to any notion of my potentially independent intelligence.[25]

At the beginning of *The Material Basis of Evolution*, Goldschmidt challenged Darwinists to come up with a plausible scenario to explain "the evolution of the following features by accumulation and selection of small mutants: hair in mammals, feathers in birds, segmentation of arthropods and vertebrates, the transformation of the gill arches in phylogeny including the aortic arches, muscles, nerves, etc.; further, teeth, shells of mollusks, ectoskeletons, compound eyes, blood circulation, alternation of generations, statocysts, ambulacral system of echinoderms, pedicellaria of the same, cnidocysts, poison apparatus of snakes, whalebone, and, finally, primary chemical differences like hemoglobin vs. hemocyanin, etc."[26] The short answer is that the Darwinists have failed his challenge miserably. They posit something within the realm of the imaginable, like 5 percent of a potential eye structure arising by chance. The problem is that 5 percent of an eye does not equal 5 percent of vision, any more than jumping 5 percent of the way across a canyon is 5 percent as good as jumping all the way across. The upshot is that there is no reason for nature to select the promising 5 percent. Natural selection, after all—especially given the materialist view of nature—does not have foresight.

Darwin, understanding the need to begin with something that could contribute at least a modicum of functional advantage, suggested the evolutionary development of the eye began with a nerve sensitive to light. He did not know how any sensitive nerve might come about, but just took it for granted that it did. From there he speculated that it could be made sensitive to light and then went on to point out optical organs in various animals and invited his readers to believe that natural selection could fill in the transitions.[27] But even a light-sensing nerve, we now know, requires considerable molecular biological complexity. And Darwinists have utterly failed to discover, or even imagine, a detailed functional pathway from one proposed stage to the next leading to the vertebrate eye. And as with so many things in biology, they aren't inching toward the goalposts. Instead, the goalposts are moving rapidly away from them as more and more is discovered regarding the bioengineering sophistication of our system of vision, which necessarily includes the eye but involves much more. The reason: These discoveries tell us that the number of simultaneous mutations needed to achieve any one functional leap forward along an evolutionary path from "light sensitive patch" to mammalian vision is much, much greater than previously imagined. The required leaps, in other words, have grown implausibly big—or to be precise, they have grown even more implausibly big than they were before.

Calculating the Odds

Darwin's theory of evolution focuses on the diversification of life since the origin of the first life. But Darwin privately hypothesized about the origin of the first living cell along with all its accompanying building blocks. Both the origin of life and its diversification are, in his view, the products of chance, which requires time, if it is to be credible. The time available for chance to do its work is no more than the age of the universe, and likely much less. In Darwin's time, there was no way to calculate the chances of a blind process giving rise to life or to its diversification. Not only did scientists lack the first inkling of the cell's complexity, but no one even knew what the essential biological elements and compounds were and how to quantify them. One cannot, for example, assess the chances of drawing a red marble out of a bag until one has some idea of how many marbles are in the bag and how many of them are red.

In the mid-twentieth century, the situation began to change. The DNA macromolecule was identified by James Watson and Francis Crick in 1953. By 1966 the genetic code was deciphered by Marshall Nirenberg and other scientists. But even before that, there was a consensus that sections of DNA, called genes, coded for proteins. Within the gene, each section of three base pairs specifies a distinct amino acid. So the gene as a whole is translated into a long chain of amino acids called a protein.

There are many hundreds of different kinds of proteins, each with a distinct role in much the way that distinct tools and distinct parts in a mechanic's shop each have a distinct role, with the difference being that protein "machines" tend to be far cleverer and more capable than the dead tools and dead parts of the mechanic's shop. These many different kinds of proteins are what build up our bodies and keep them running through many diverse processes: metabolism, breathing, seeing, hearing, healing, etc. The various proteins are made according to the information encoded in DNA, which itself needs to be reproduced whenever a cell reproduces. Every replication process known to man involves occasional errors. DNA's replication process is astonishingly accurate, but it too makes the occasional error. It was thought that these errors, these random mutations in DNA, would give rise to mutations in proteins, with a tiny minority of these, it was supposed, providing an advantage to the host organism. This was the supposed source of the mutations that Darwin needed for his theory.

The problem for Darwinism was that the newfound level of knowledge about DNA copying errors allowed some basic mathematical calculations, and the results were not friendly to the theory. There are twenty primary amino acids. If you are talking about an amino acid chain one link long, there are twenty possibilities. If two amino acid lengths, then 20 x 20 possibilities—400. Three amino acid lengths, then 20 x 20 x 20—8,000. And so on. It is easy to see that the total number of different proteins that could be assembled from an amino acid chain 200 links long is 20^{200} or roughly 10^{260}. This should strike everyone as a very large number. If it were written out, it would be the number 1 followed by 260 zeroes. In comparison, the total number of elementary particles in the universe is about 10^{80}.

The question that follows is this: How many of those possible combinations make for functional, foldable proteins, and how many make for useless gibberish? If it's, say, half and half, then we could imagine DNA mutations hitting on new and beneficial protein types without much trouble, and evolving step by step. But what if the functional sequences are dwarfed by the dysfunctional ones, as for instance in language. Randomly bang away on a keyboard till you have 200 characters. What are the odds that the resulting text will be a coherent passage of words building into meaningful sentences? Practically zero, and that's because the percentage of gibberish sequences absolutely dwarfs the percentage of meaningful ones. But are amino acid chains like human language in this regard?

Murray Eden, a professor of engineering and computer science at MIT, took an interest in this problem in the early 1960s. He was especially skeptical as to what slight changes in the coding gene would do to the resultant protein. He knew that the slightest mistake in a formal code, such as a computer program, could destroy it. So it seemed to him that mutations are not the place to go and look for viable changes. Other mathematicians agreed, much to the consternation of the Darwinists.[28]

A conference took place at the Wistar Institute in Philadelphia in 1966 between some very prominent thinkers from mathematics and biology. In addition to Eden, among the concerned mathematicians were Marcel-Paul Schützenberger and Stanislaw Ulam, the co-designer of the hydrogen bomb. The Darwinists present included Ernst Mayr, Richard Lewontin, and Peter Medawar, who chaired the meeting.

It was a lively gathering. As Paul Nelson says, "The official monograph that followed the conference—*Mathematical Challenges to the Neo-Darwinian Interpretation of Evolution* (1967)—features transcripts of the conversations and one can all but hear the attendees tossing chairs at each other."[29] The Darwinists insisted that the mathematicians must be wrong because Darwinism is true. The biologist C. H. Waddington accused Schützenberger of arguing for special creation, to which Schützenberger and others in the background shouted "no." But they could not offer any other solution as to how life originated and diversified.[30]

The only way to escape the seemingly astronomical odds against Darwinism was to hope that most proteins would be functional. Another illustration might be helpful. We are familiar with electronic locks in large office buildings. A company might give each employee a unique 6-digit code to open the door. Such systems are designed to let the administrators know who has entered the building. The total number of possibilities for a six-digit code is one million, and no thief who has any notion of probability would waste his time trying to get in if there was only one number to open the lock. But if the building has 10,000 employees, each with a unique functional code, then the chances of getting into the building improve to one in a hundred, which would make it relatively easy for a thief with a bit of patience to enter.

When it comes to proteins, Darwinists hoped that many different chains of amino acids could do something useful. If most could do something useful, or even a sizable fraction of the total could do so, then the astronomically high number of variants need not be a roadblock to evolution. Random mutation would not need to stumble across the needle-in-a-haystack functional protein. The functional sequences would be so common as to be hard to miss.

In 1966, it was hard to get a handle on the ratio of functional proteins to the total number of possibilities. In the years since, several molecular studies have been carried out to get some idea of this ratio. The results do not look good for the Darwinists. Douglas Axe, with a PhD from Caltech and a postdoctoral research stint at Cambridge to his credit, is one such researcher. Proteins are long chains of amino acids, but by and large they will not be of any use unless they fold into compact shapes that enable them to perform highly specific functions. Unfolded proteins tend to get chewed up and recycled. Axe studied protein folding through meticulous laboratory research and published the findings in a leading science journal. He estimated that the ratio of sequences that correspond to a functional, foldable domain in a relatively simple protein versus all sequences of the length of the domain (153 amino acids) is one in 10^{77}.[31] A domain is a relatively independent portion of a protein. Later research confirmed that many proteins correspond to extremely rare functional sequences.[32] Given that the total number of organisms that have lived on the earth from the beginning is only about 10^{40}, the likelihood of stumbling across most functional proteins by chance is, for all practical purposes, nil.[33]

Dawkins has tried to deal with the mathematical problem by running a computer program to beat the odds. He chose a phrase from Shakespeare's *Hamlet*, "Methinks it is like a weasel," and programmed a computer to generate it randomly. Since there are twenty-eight positions in this phrase and the number of letters in the English alphabet is twenty-six, plus a twenty-seventh for a space, then the chances that a monkey randomly typing away will produce the required phrase (leaving out capitals and punctuation marks) is one in 27^{28} or about one chance in 10^{40}. So how did Dawkins's computer get to the required phrase so quickly? He got his program to produce random strings of twenty-eight letters and spaces. The program then chose strings that resembled the target more closely than others and proceeded to the next generation. And, sure enough, the computer found the phrase in about half an hour.

But is this proof of natural selection? Clearly not, because unlike Dawkins, natural selection has no goal in mind. Natural selection as Darwinists understand it is blind—God is not steering it behind the scenes; Dawkins knew the outcome that he was trying to reach and made sure that his machine would get there. He acknowledged as much, but then one wonders why he would bother running the simulation.[34] One suspects it was to give some level of confidence that it was possible to overcome the odds stacked against Darwin's theory. But if foresight is taken out of the program, the astronomical odds reappear with unabated strength.

A common move at this stage is to first admit that the odds against life arising by chance are slim indeed but that these odds can be overcome because the number of tries at first life is vastly greater than one might imagine, even infinitely greater. Fred Hoyle argued this by pressing the idea that the universe is infinitely old,[35] but that idea was discredited as the evidence for a cosmic beginning became overwhelming in the latter half of the twentieth century.

But there is another way to greatly expand the probabilistic resources for lucking into first life. The bioinformatician Eugene Koonin calculated the odds for the chance origin of life at less than 1 in $10^{1,018}$—a number so large that it makes the number of elementary particles in the universe seem like the number of marbles you could fit into a kid's hand by comparison.[36] Koonin then explains away the problem by invoking an infinity of universes, with ours being one of the lucky ones that won the "origin of life" dice game. He invokes this solution not because there is independent evidence for this infinity of universes. It's rather that materialism demands it. On this bit of creative accounting, we won the cosmic lottery because someone has to win when there is an infinity of universes. At this point, anyone with a modicum of common sense should know that we are dealing with a materialist *credo* rather than empirical science.[37]

Irreducible Complexity

The field of molecular biology might intimidate the non-specialist into believing everything that the Darwinian sages proclaim. So it is no surprise that Lehigh University biologist Michael Behe caused quite a stir when he published *Darwin's Black Box* in 1996. He revealed that the complexity of molecular biological systems and their specification is such that there is little hope that they could have arisen by the interplay of chance and necessity. He introduced a new concept into the debate: irreducible complexity. The Darwinian scenario has to go from the simple to the more complex via small stages. Only advantageous mutations will be selected. What is one to think, then, of the origin of a complex system that needs numerous very specific proteins carefully fitted into a whole before it serves any function? How would any of the intermediate stages get selected by a blind evolutionary process that has no patience for dysfunctional half-measures and quarter measures?

Behe's most famous example is the bacterial flagellum, a whip-like appendage that gives bacteria mobility. It has some very impressive properties, such as spin rates of tens of thousands of rpm and the ability to reverse direction almost instantaneously. The simplified diagram of the proteins involved looks like an engineering blueprint for an outboard motor. Moreover, a system has to exist to produce and put into place in a coordinated fashion all the different components. If any of these things are missing, the flagellum does not work. So there is no reason for a partial flagellum to be selected. The flagellum quickly became an icon of the intelligent design community.

Behe used a mousetrap to illustrate the concept of irreducible complexity. Unless there is a base, a spring, a means of attaching the spring to the base, and a trigger, the device will not catch any mice. It is irreducibly complex. As one might imagine, this stimulated Darwinist bloggers to imagine even simpler mousetraps that could serve as evolutionary intermediates on the way to the common mousetrap Behe described, but their efforts were not very convincing.

Meanwhile, Behe's flagellum example sparked more serious efforts among scientists to come up with a plausible smooth evolutionary pathway from simple precursor to full-blown bacterial flagellum. The results have not come even close to meeting Behe's challenge. In 2006, there was a paper by Mark Pallen and Nicholas Matzke in *Nature Reviews Microbiology*, which compared flagellar protein sequences in the many different bacteria that possess them and found some similarity, suggesting a common ancestral gene. They also showed that some non-flagellar proteins were somewhat similar to the flagellar ones.[38] But they did not present any explanation of how the proteins might have evolved by smooth transitions.

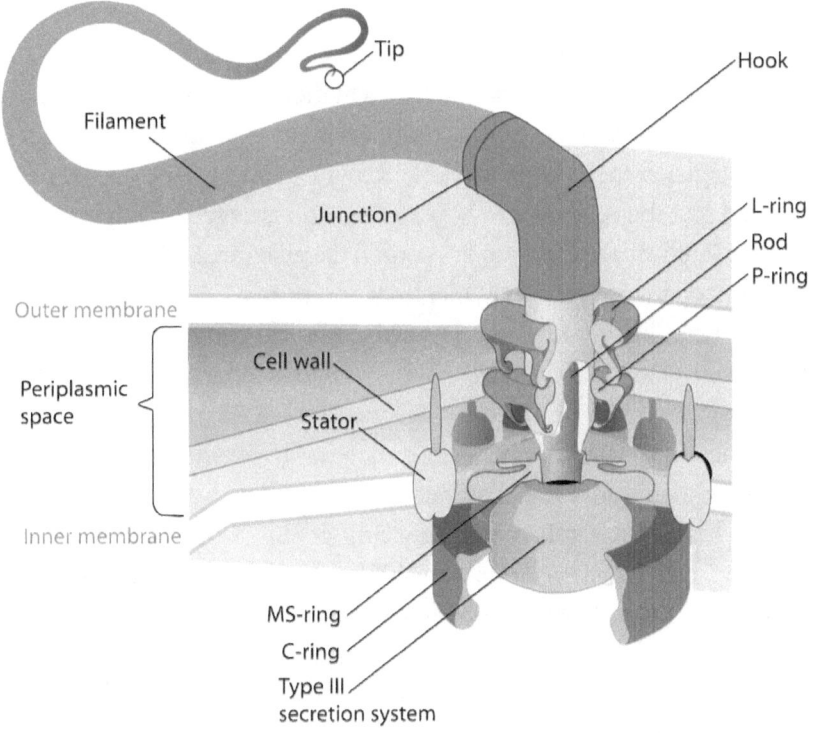

Figure 2.2.
Various parts of the bacterial flagellum.

Jonathan McLatchie wrote that the Pallen and Matzke paper "trivializes the sheer complexity and sophistication of the flagellar system—both its assembly apparatus, and its state-of-the-art design motif." McLatchie goes on to say that "the process by which the bacterial flagellum is self-assembled within the cell is so sophisticated that I have long struggled to convey it in an accessible way to lay-persons." Then, after delving into the engineering details of the system for several paragraphs, he concludes, "My description, given above, has really only scratched the surface of this spectacular item of nano-technology.... I have not, for the sake of brevity, even discussed the remarkable processes of chemotaxis, two component signal transduction circuitry, rotational switching, and the proton motive force by which the flagellum is powered.... But the bottom line is that modern Darwinian theory—as classically understood—has come nowhere close to explaining the origin of this remarkably complex and sophisticated motor engine."[39]

The next year a paper was published in the *Proceedings of the National Academy of Sciences USA* with the title "Stepwise Formation of the Bacterial Flagellar System." The authors, Renyi Liu and Howard Ochman, claimed that they could show how all the various flagellar proteins arose by gene duplication from one primordial flagellar gene. The paper came recommended by Kenneth Miller (more on him in Chapter 8).[40] And it served the political purpose of attacking the ID community, not long after a court case in Dover, Pennsylvania had given it prominence. But the paper itself was attacked by Matzke, no friend of ID, who wrote that much of the paper "ranges from dubious to irremediably wrong."[41]

Behe also addressed a paper entitled "Evolutionary Resurrection of Flagellar Motility via Rewiring of the Nitrogen Regulation System," published in *Science* in 2015. The authors turned off the master control gene while leaving the dozens of other flagellar genes intact. A few days later, the maimed bacteria were swimming again, to the cries of "giant evolutionary leap" from *The Scientist*. Behe was less impressed. The system was turned back on but at the expense of nitrogen regulation in the swimming bacteria. Behe suggested that a more honest title would have been "Bacteria Endure Crippling Mutation to Stave Off Extinction."[42]

It should be clear that no one has proposed a smooth stepwise path for the evolutionary development of the flagellum. Not even close. Not even in the same zip code, area code, or continent. And the goalposts are moving farther away, not nearer, as molecular biologists continue to uncover new levels of complexity in this molecular machine.

And there are, of course, many other such molecular systems. In *Darwin's Black Box*, in addition to the flagellum, Behe discussed the blood-clotting system, vesicular transport, and the immune system. All are fascinating and complex. And Behe does a thorough job of explaining them to the lay reader to give some sense of the irreducible complexity involved in each of them. Even Behe's opponents concede that he is on to something. In 2001, Franklin Harold published *The Way of the Cell*, with this telling admission: "We should reject, as a matter of principle, the substitution of intelligent design for the dialogue of chance and necessity (Behe 1996); but we must concede that there are presently no detailed Darwinian accounts of the evolution of any biochemical system, only a variety of wishful speculations."[43] This state of affairs remains true more than two decades later.

The argument that is perhaps the clearest and most accessible to all readers of *Darwin's Black Box* is Behe's analysis of biochemical textbooks and other literature. He provides a list of thirty popular university-level biochemical textbooks, published between 1970 and 1992. None of them has fewer than 1,000 entries in the index; most have 5,000; and one has 10,000. Thirteen have no index entry under "evolution." The most that any one has is twenty-two. And the majority of these refer the reader to the introduction in which the author invites students to consider all biochemistry in light of evolution. Other entries take the reader to some discussion of a particularly noteworthy biological feature. The authors then credit this feature to evolution without in any way suggesting how it may have evolved.[44] Substantive attempts to make a solid evidential case for Darwinian evolution are conspicuous by their absence.

"Junk" DNA

Much has happened since the publication of *Darwin's Black Box* in 1996. But most of the new research presents challenges to Darwin's theory. The term "junk DNA" was thrust into the scientific lexicon in 1972 by the geneticist Susumu Ohno.[45] The term applied to the majority of base pairs in a genome that did not code for proteins, roughly 98 percent in the case of humans. Many thought that "junk DNA" was a powerful argument for the Darwinian story. Life, they argued, is cobbled from the 2 percent, whereas the rest is largely accumulated junk from millions of years of evolution. Intelligent design theorists, meanwhile, predicted that much of this so-called "junk DNA" would turn out to have important function.[46]

With the development of genome sequencing and its widespread applications to animals ranging from bacteria to man, the percentage of the genome that codes for proteins has not grown, but less and less of the non-coding DNA is looking like junk. Instead, many functions have been discovered for the supposed junk, such as transcriptional and translational regulation of protein-coding sequences and much more besides. The Encyclopedia of DNA Elements (ENCODE) estimates that 80 percent of human DNA may be functional—that is the percentage of the DNA that is actually transcribed.[47]

It may be that the amount of "junk" DNA is only a measure of our ignorance of it. As Francis Collins, once an enthusiast for the term, put it in 2015: "In terms of junk DNA, we don't use that term anymore because I think it was pretty much a case of hubris to imagine that we could dispense with any part of the genome, as if we knew enough to say it wasn't functional…. Most of the genome that we used to think was there for spacer turns out to be doing stuff."[48]

Clearly, then, the evidence on this front is breaking in favor of intelligent design.

Homologous Genes

Another significant discovery that clearly bears on evolution involves evolutionary developmental biology, or *evo-devo* for short. As Michael Denton puts it:

> At the heart of these advances has been the discovery that a limited set of highly conserved genes, gene circuits, and developmental mechanisms—*Hox* genes, signalling proteins such as 'sonic hedgehog,' chemical gradients, and gene regulatory networks…—are involved in the construction of the bodies of all bilaterally symmetric animals, all the morphological homologs, and indeed of all higher organismic form. Sean Carroll has termed this set of conserved genes and developmental mechanisms the "toolkit."[49]

To put this into more vivid form, there is a gene called *Pax6* which is found in both humans and in fruit flies, whose mutation in each species causes maldevelopment of the eye. Denton continues: "It is now known that *Pax6* turns on the genetic circuitry which leads to the development of the eye in a vast range of animals, even though the eyes, such as those of insects, vertebrates, and cephalopods (squid) could hardly be more different in fundamental design."[50] Giuseppe

Sermonti writes of the Hox cluster of genes: "All animals have the same cluster of genes dictating the order in their body regions. The cluster, therefore, was universal and of great antiquity, since it belonged to the progenitor of all animals and had been preserved intact throughout the development of the fauna for 500 million years. The genes decreeing the general organization of a mammal are the same as those decreeing the organization of an insect."[51]

The Darwinist propaganda machine makes much of this as favoring common descent, but Darwinism did not at all expect this finding. Darwinists thought the molecules of life would be mutated so often on evolution's path from bacteria to animals that there would remain little if any trace of ancient genes. Thus, Ernst Mayr, a key player in the neo-Darwinian synthesis, claimed that "the search for homologous genes is quite futile except in closely related species."[52] Well, that Darwinian prediction turned out to be way off.

Epigenetics

There was a great shift in biology with the discovery of DNA and the genetic code. The science could go indoors into the lab. Its technical powers greatly expanded. Moreover, mutations in DNA were latched onto as the driving force of evolution, despite the astronomical odds against finding functional proteins by chance. The focus on DNA has led many people to speak of DNA as the blueprint for life. But is this true?

There are several considerations that should make it clear that DNA is far from the whole story. Even if one acknowledges that DNA does much more than code for proteins and that there are codes within it to control various molecular processes, one runs into undeniable empirical evidence that there is more to the specification of plants and animals. This "more" is usually referred to as epigenetics, from the Greek word "epi" meaning "over" or "above." Various laboratory tests suggest that something other than DNA is operative in the growth of new life. In the 1930s and 1940s, Ethel Harvey removed the nuclear DNA from sea urchin embryos, which went on to develop up to 500 cells before stopping. In the 1960s Belgian scientists chemically blocked transcription of DNA into RNA in embryonic amphibians, which went on to grow to several thousand cells before stopping. And in the 1970s, Canadian scientists removed the nuclear DNA from frog embryos but injected the cell-division apparatus from a sea urchin. The embryos were able to make good progress, although not to the end.[53]

But we need not go to such technical lengths to arrive at the same conclusion. Giuseppe Sermonti points out some obvious truths that should give anyone who thinks that DNA is the whole story many reasons to reconsider. For example, the caterpillar and the butterfly have exactly the same DNA, yet their form is very different. All the termites in a colony have the same DNA, yet there is a great difference among the king and queen and the workers. "In many species the difference between male and female is a matter of chromosomes, yet the sexes may be scarcely distinguishable," he writes. "In some species, however, male and female have the same chromosomes (the same DNA) yet they are spectacularly different."[54]

Nature provides many examples of the same forms but with significantly different DNA. The most obvious cases come from comparing marsupial and placental mammals. Sermonti writes, "A gerbil hops about on long hind legs, its small forelimbs clasped to its chest, as it disappears into the grass. I defy anyone to determine by sight whether it was a marsupial or a placental mammal."[55]

Sermonti then summarizes:

> Between DNA and an organism's appearance lurks the black box that conceals within itself the entire mystery of typologies and differences among organisms. The discipline that explores this *terra incognita* has been called "epigenetics." There is something infelicitous about the name, which still implies the theory that DNA, via a series of operations and interactions, when acting within a specific context, ends up producing an organism. Although it posits a distance between DNA and the final appearance of an organism, between genetics and form, between the genotype and phenotype, epigenetics is still about how DNA sets about "making" an organism. My view is that the problem is rather how the organism makes use of DNA, getting it to work or keeping it silent, and how it selects its areas of interest. DNA is not the starting point.[56]

Michael Levin, the director of the Allen Discovery Center and Associate Faculty at Harvard University's Wyss Institute, paints a similar picture. He makes the obligatory Darwinian nods, but he admits that no one has any clear idea how biological shapes come about:

> While genomes predictably encode the proteins present in cells, a simple molecular parts list does not tell us enough about the anatomical layout or regenerative potential of the body that the cells will work to construct. Genomes are not a blueprint for anatomy, and genome editing is fundamentally limited by the fact that it's very hard to infer which genes to tweak, and how, to achieve desired complex anatomical outcomes. Similarly, stem cells generate the building blocks of organs, but the ability to organize specific cell types into a working human hand or eye has been and will be beyond the grasp of direct manipulation for a very long time.[57]

In his book *How Life Works*, Philip Ball, who was one of the editors of *Nature* for twenty years and who considers himself a Darwinian, also denies that the genome is a blueprint for life:

> Is the genome *informationally complete* to specify an organism (as we might expect of an instruction book or a blueprint)? The answer to that is an unequivocal "No." You can't compute from the genome how an organism will turn out, not even in principle. There is plenty that happens during development that is not hardwired by genes. And from a single protein-coding gene, you can't even tell in general what the product of its expression will be, let alone what function that product will serve in the cell. You couldn't even predict that the genome *will* build a cell—for the simple reason that a genome can't build a cell. Rather the cell is a precondition for anything that the genome can do. Put a full set of chromosomes into water, perhaps with all the ingredients our bodies ingest to make its biomolecules, and nothing of note will happen. We can be as generous as we like with such provisions, for example also adding the lipid molecules that make up cell membranes, and premade enzymes for transcribing and translating, and still it will be to no avail. An organism won't spontaneously form. For such reasons alone, a genome sequence can never "fully specify the organism": how it will grow, how it will look, how it will behave—or in short, how it works.[58]

The eminent biologist Denis Noble, who has been trying to set Richard Dawkins straight on what genes can and cannot do, wrote a laudatory review of Ball's book in *Nature*. "Ball is not alone in calling for a drastic rethink of how scientists discuss biology. There has been a flurry of publications in the past year [2023], written by me and others. All outline reasons to redefine what genes do. All highlight the physiological processes by which organisms control their genomes."[59] Biology needs to free itself from the constricting and erroneous dogma that we are our genes. "Epigenetics" is perhaps a clumsy but necessary way to describe the future of biology.

Much of this field is unexplored. But one thing that should be clear is that it presents yet another level of formidable odds against Darwinism. Axe's calculations of hitting a functional protein by chance already point out the implausibility of the Darwinian paradigm. The odds of assembling a primitive "toolkit," discussed above, just exacerbate the problem. Finally, epigenetics makes clear that, generally speaking, the unit of selection cannot be less than a functional cell, which contains the genome and a structured environment.

Complex Behaviors

Many animals exhibit amazingly complex behavior, in many cases instinctively. Salmon, migratory birds, sea turtles, and monarch butterflies are masters of navigation. So are many species of ants. A foraging ant, once it has found something, will take the shortest path back to the colony. Termites, bees, and ants are master builders. Compass termites in Australia orient their nests so that the slim side will face north. This prevents excessive heating from the noonday sun. Other termites open and close pores in the nest to keep temperature and carbon dioxide levels relatively constant. Honeybees can keep their hive between 93 and 96 degrees F, despite external temperatures varying from -20 to 120 degrees F. And both ants and termites have been practicing agriculture for millions of years. There are no plausible Darwinian scenarios to explain these complex behaviors from these tiny-brained creatures. They seem to be operating based on highly sophisticated in-built algorithms, not at all the sort of thing that lends itself to being built up one tiny functional mutation at a time over long ages, as the Darwinian paradigm demands.[60]

Insect Galls—Thousands of Reasons Darwin Was Wrong

In *The Origin of Species*, Darwin wrote: "If it could be proved that any part of the structure of any one species had been formed for the exclusive good of another species, it would annihilate my theory, for such could not have been produced through natural selection."[61] Wolf-Ekkehard Lönnig has written a paper with the subtitle "How More Than Twelve Thousand Ugly Facts are Slaying a Beautiful Hypothesis: Darwinism."[62] As he notes, the number could be much larger, since naturalists estimate the number of different gall-inducing insects to be between 21,000 and 211,000, yet 12,000 is a nice conservative number to make the point.

What are galls? Here is Marion Harris and Andrea Pitzsche's definition: "A gall is a manifestation of the reprogramming of plant cellular growth and/or development… that begins at the colonization site of a specific foreign organism, which receives specialized services from the plant and continues to interact with the *de novo* plant tissue or organ as it develops and matures."[63] Perhaps some pictures can help make sense of this definition. They certainly explain why Lönnig spoke of "Ugly Facts."

Galls are caused by insects laying their eggs on a plant. What does the insect get? A protective housing for the eggs and larva and eventually food for the larva. What does the plant get? Nothing. In another article, Lönnig cites Joachim Illies: "For the plant, the entire effort involved in the gall formation is of no apparent benefit, it is more of a harm because it requires nutrients, reduces the assimilating leaf area and disrupts the normal course of growth, sometimes even the most valuable parts of the plants: buds and seeds. Consequently, according to Darwin, the plants without galls should have an advantage over those with galls, and so in the course of evolution the gall-free variants among the plants should have been chosen very soon and everywhere as the fittest ones." But that isn't the case.[64] So next time that you see

Figure 2.3.
Lime nail galls caused by the mite *Eriophyes tiliae*.

these ugly growths, recall that on Darwin's own accounting, they "annihilate" his theory.

How Far Does the Evidence for Transformation Take Us?

I hope that by now it is clear that there is no real empirical evidence for the grand claims of Darwinism: namely, that all life originated from a single-celled individual by a blind process of random variation and natural selection. But does that mean that there is no evidence for evolution? A somewhat involved answer is required here because of the different meanings of evolution.

There is substantial evidence for the power of natural selection to fine-tune species. Darwin's finches are an icon of evolution for a good reason. There are perhaps fourteen species of finches in the Galápagos Islands. (Taxonomists incessantly debate the exact number.) It is thought that an original population of birds was blown from mainland South America to the archipelago soon after the formation of the volcanic islands some three to five million years ago. These adapted to the different niches found on the islands and vary in the size of their bodies and the size and shape of their beaks. Thick and stubby beaks help one species break open seeds. Long and thin beaks make it possible for another to feed on nectar.

Peter and Rosemary Grant have studied the finches since 1973. In 1977 there was a severe drought on the island. The only seeds available were large and hard to crack. As a result only the larger finches with bigger beaks survived. The mortality rate among medium ground finches was 85 percent. The Grants measured the survivors and found that on the average they were 5 percent larger than the average previous to the drought and the next generation too was bigger. They caught natural selection in action. And over the decades they were able to show that the body and beak size of the medium ground finch tracked the weather.

But there are limits. Finches still stay finches. And both the Grants and other biologists visiting the Islands have noted that some "species" interbreed with other "species" and produce viable and even more vigorous offspring than the pure breeds. Since modern taxonomy uses the biological species concept to differentiate species—species are groups of similar individuals that breed in natural circumstances and produce fertile offspring—these interbreeding episodes lead some taxonomists to lump the six different species of medium ground finches into one. The authors of one such study have coined the term "Sisyphean evolution."[65] In the course of a million years, incipient speciation was frustrated time and again by interbreeding to keep the finch the finch.

The Grants also studied the finch genomes to get a sense of what was going on. They determined that the blunt beaks were most likely the result of a mutation in a single gene called *ALX1*, which, in humans, can cause severe maldevelopment of the head and face. The protein is 326 amino acids long and the birds with the blunt beaks have a mutation in two of the 326 amino acids. Computer simulation has deemed the mutation damaging. But, in this case, the "damage" allows the exploitation of bigger and tougher seeds, so it is useful and selected for. It is thought that this mutation arose soon after the finches arrived. No other variation of this gene was found. So over millions of years, the two versions of the gene helped finches survive in the changing climate. The finches of the Galapagos clearly show variation and may be classified as different species. But they remain finches.[66]

A more spectacular example of evolution are the cichlid fishes of Lake Victoria in Africa. As Behe reports, seventeen thousand years ago, Lake Victoria dried up and stayed that way for about two thousand years. When wet weather reestablished the lake, fertilized fish eggs carried on the feet of wading birds, or perhaps intentional human activity, introduced fish into the lake. Behe writes: "Until the last few decades, during which disastrous wildlife-management decisions decimated their numbers, the lake was host to about five hundred species of fish that are found nowhere else in the world, and since there was no lake until relatively recently, that means those species must have evolved in place in just the past fifteen thousand years!"[67] These fish range from a few inches to six feet. They have a wide variety of colors, which make them attractive to aquarium buffs. The tilapia is a gourmet favorite. Nevertheless, George Barlow, the ichthyologist and author of *The Cichlid Fishes* (2000) said that "what seem to be major changes in appearance have evolved with little alteration of the basic plan."[68]

The genomes of these fishes are still being investigated to see what might have happened.[69] But one very interesting fact is clear, which nicely summarizes the potential of a natural process to create new species. Behe writes:

> The cichlids of Lake Victoria evolved in the last fifteen thousand years or so—the time that the current replenished lake has been in existence. Yet Lake Malawi is over a million years old, and Lake Tanganyika about 10 million. Despite the vast differences in age, all have roughly the same number of cichlid species. What's more, the independently evolved lineages of each lake often resemble each other closely, clearly demonstrating the limited range of available

variation, which apparently can appear very rapidly—and then just as quickly stagnate.[70]

All in all, there are about 1,500 species of cichlids on earth, comprised under seventy-five genera. But none of these is so different from any other as to fit into a new family.

The finches and the cichlids have been extensively studied, but there are other plants and animals that reveal the same kind of limited variety. Behe provides a list of "luxuriantly evolving groups," which in addition to the finches and cichlids includes but is not limited to 300 species of anoles (iguanian lizards) in three genera, fifty-five species of honeycreepers (birds) in twenty-four genera, and 1,000 species of fruit flies in two genera.

This leads Behe to posit that the limit to evolution by natural selection is set at the level of family, in the Linnaean system. One can, of course, get into a never-ending argument about the reality of any classification system, but it is easy to see what Behe is driving at. The changes that natural selection can attain are such that the animal or plant is still recognized as something not too far away from its ancestors—both recent and distant.

The varieties of cichlids have become so well known for their quick evolution that they have made it into the "Science and Technology" section of the *Economist*. Genetic analysis has shown that hybridization is the driving force behind the change, which, the article admits, would have been considered a heretical hypothesis not too long ago. Hybridization was thought to be a dead end because many hybrids, such as mules, turned out to be infertile. And in fact there was a law—Haldane's rule—that was supposed to explain how natural selection set up barriers against hybridization. It now turns out that hybridization is of great importance. The cichlids of Lake Victoria are thought to be hybrids of two populations, one from the Congo River and one from the Nile. The *Economist* article also mentions birds of the genus *Sporophila* as hybrid productions, and many other cases involving smaller varieties of hybridization. The mixing of genes from two similar populations gives natural selection lots of material to work on. There is no need to wait interminably for the right random mutation.[71]

Evolution as Degradation

Such instances are, indeed, evidence for evolution, but notice that they are all cases of relatively modest evolutionary changes within a species or among closely related populations.

Moreover, as Behe shows, most often the genetic change involved in the evolution of finches, cichlids, and other animals is really a degradation. He makes the case in his book *Darwin Devolves*, where he provides numerous examples. For example, the polar bear is thought to have diverged from the brown bear several hundred thousand years ago. The polar bear has a mutated gene APOB that allows the animal to survive on a very high-fat diet. This is good for an animal whose main food is seals. In humans and mice, a mutated form of this gene causes high cholesterol and heart disease. The polar bear also has a mutated form of the gene LYST, which is probably associated with the white pigmentation. A computer analysis of the mutant variety revealed that the changes were of themselves damaging to protein function. In the case of the polar bear, there were seventeen mutant genes that were highly selected; only three to six of these were free of degrading functionality.[72]

The degradation of genes is also what has been observed in Richard Lenski's long-running experiment in evolution on *E. coli* bacteria. Started in 1988, the experiment reached 78,000 generations, when it was frozen at the beginning of the Covid pandemic. (It has been restarted since then.[73]) There have been some interesting developments. The most publicized "positive gain" was a strain of the bacteria that developed the ability to metabolize citrate in the presence of oxygen. This may sound impressive, but the bacteria already had the ability to "eat" citrate. As Matti Leisola and Jonathan Witt put it, "The lab eventually determined that a gene, *CiT*, which encodes a protein that normally imports citrate into the cell when oxygen is *not* present, had mutated. The mutation gave the protein the ability to import citrate. Notice that the protein already had the ability to import citrate."[74] So only a switch that prevented the transport when a toxic condition (i.e., oxygen) was present, broke down. Nothing new was created.

Behe points out that the Lenski experiment, far from showing how new complex genes arise, shows the destructive power of Darwinian evolution. The degraded genes, such as the switch that made it possible for *E. coli* to ingest citrate, are forced by natural selection to be fixed into the population. Natural selection has no foresight; so it favors whatever might be advantageous in the instant, even if it comes with long-term disadvantages. As Behe puts it, "Those initial random 'beneficial' citrate mutations that had been seized on by natural selection tens of thousands of generations earlier had led to a death spiral."[75]

Behe provides a vivid picture of how to think about such degradations that can in some circumstances be helpful. Imagine you have a car and need to drive it as far as possible on a tank of gas. Designing a whole new kind of energy-efficient engine is one way to do this. But if you are in a hurry—in danger of death unless you can go 500 miles with no chance of getting more gas along the way—your best bet is to make the car lighter. Throw away the jack and spare tire. Jettison the air-conditioning compressor or at least disconnect it. Throw out the seats. Great. You made it. But your car is not going to be able to do some of the things that people want in cars. So it is with biology. An animal can adjust to fit a niche. But its adjustments often leave it stranded, with less potential to go elsewhere if conditions change.

One more example. In *The Edge of Evolution*, Behe analyzed the relative frequency of sickle cell anemia in human populations in Africa where malaria is present. Of itself, the gene that leads to sickle cell anemia is a degradation, obviously so. But it is a very helpful degradation to have for surviving malaria. No other explanation besides natural selection need be invoked to explain the prevalence of this degradation in various African populations, who live in regions where malaria is a constant problem.

Behe is not alone in positing that gene degradation is responsible for Darwinian successes. A 2021 article in *Nature Heredity* notes:

> Views on loss-of-function mutations—those abolishing a gene's biomolecular activity—have changed considerably over the last half century. Early theories of molecular evolution that emerged

during the 1960s and 1970s saw little potential for loss-of-function mutations to contribute to adaptation. Except in the case of inactivated gene duplicates, nonfunctional alleles were often assumed to be lethal, with adaptation being generally regarded as a process explained only by the fixation of single, mutationally rare alleles that improved or altered a gene's function.... Discoveries during the subsequent two decades have continued to support the idea that loss of function contributes to adaptation, with cases of adaptive or beneficial loss of function being discovered across diverse organisms, genes, traits, and environments.[76]

The article duly cites earlier research in the field, notably a paper from 2020. Conspicuous by its absence is a citation to Michael Behe, who had published similar work in 2010.[77] But, as we will see in the next chapter, radio silence is the strategy when dealing with the work of proponents of intelligent design.

Darwin: Partly Right

Clearly, as Behe acknowledges, natural selection plays a limited role in evolution. Darwin is partially right; there is no need to deny him that insight. But the evidence for the grand theory is not there. People who doubt the grand theory need not be caricatured as requiring a special creation of each different kind of finch or cichlid. Nor will they deny that bacteria can, for instance, evolve antibiotic resistance.

During the seminar I took on the history of evolutionary biology, it occurred to me—as it must have occurred to many others—that Darwin's theory can be compared to Newton's physics. For (relatively) low-speed macroscopic interactions—from cannon ball trajectories to bridge building—Newton's equations are more than sufficient; and they are much easier to handle than general relativity or quantum mechanics. But Newton's analysis could never have predicted the Big Bang or black holes; nor could it be of any use in understanding chemistry or in developing transistors. Newton's physics is fine in its limited field; and so is natural selection. Insofar as it culls the weak, it is an important element in the survival of a species. But it is not up to explaining the arrival of new orders or phyla.

Common Descent or Common Design?

The evidence for common descent of all life-forms is stronger than is the evidence for Darwin's joint mechanism of random variation and natural selection as the driving agent of evolution. Stronger, though not conclusive. One reason it falls short of conclusive is that the mechanism—if it exists—for large-scale changes is unknown. If one is open to divine causation, then the problem of a missing mechanism disappears. At the same time, if one is open to divine causation, one is not compelled to accept some one ancient organism as the ancestor of all life today. Nature may not make any leaps, but the Creator of the universe certainly can. One should also be aware that arguments for common descent do not necessarily prove that everything came from one primordial cell (universal common descent). Perhaps each of the various phyla were created independently, and then intelligently guided evolution built up the variety within each phylum.

Some people find universal common descent aesthetically very pleasing. Does it not give God more glory that He should bring out birds from dinosaurs than that He should let the dinosaurs die out and then create birds *ex nihilo* or out of clods of dirt? This is an aesthetic and theological argument. I would only caution against presuming too readily to know how God should have done something.

All forms of life—plants, bacteria, animals—contain DNA. The genetic language, or "genetic code," for translating the information stored in DNA into the amino acid sequences of proteins is nearly universal throughout all living organisms. There are many variations on it,[78] but we can still ask, why is it very nearly universal? There are multiple possible explanations. One possibility is that the code is simply arbitrary, and the near universality suggests inheritance from one generation to the next throughout the history of life. Under this view, the best explanation for the near-universality of the genetic code boils down to the common ancestry of all life. But there's another way to explain this evidence: there could be functional reasons for re-using the same genetic code over and over again in different organisms. This is also called "common design." It turns out there is good evidence backing the "common design" view. The genetic code appears optimized to reduce the deleterious effects of mutations.[79] The genetic code of life is an ideal code to use across organisms to minimize harmful mutations, and there are rational reasons why a designer might re-use such a code throughout all life.

The question of common ancestry is often complex. Some discoveries associated with evo-devo might suggest universal common descent. For example, the same gene toolkits are found in greatly diverse life-forms. The *PAX6* gene is involved in the development of the eyes in animals ranging from fruit flies through squid to humans. But there's more to it. The supposed common ancestor of these diverse organisms isn't even known to have had eyes, making it difficult to explain the re-use of *PAX6* to control eye development through inheritance from some common ancestor with eyes. Rather, evolutionists must view the re-use of *PAX6* as a form of genetic "convergent evolution"—where widely different organisms independently evolve the same trait. Such cases appear unlikely to occur on Darwinian grounds but are readily explained by common design.

Others may cite evidence that classification schemes based on molecular similarity sometimes line up with Linnaean classifications based on phenotype. Originally, in the pre-molecular era of biology, phylogenetic trees were constructed by comparing the anatomical or morphological traits of living and extinct (fossil) organisms. But since the revolution in gene sequencing of the past few decades, we can now compare genomes and construct phylogenetic trees showing the evolutionary relationships of organisms based upon similarities and differences in their molecules—i.e., their DNA sequences. If the molecule-based trees turned out to corroborate the morphology-based trees, this would suggest that common ancestry, as suggested in the tree of life, is real. But that isn't what happened.

These two lines of research (morphological and molecular) are not converging on a unique tree of life. Casey Luskin explains:

> A June 2012 article in *Nature* reported that short strands of RNA called microRNAs "are tearing apart traditional ideas about the animal family tree." Dartmouth biologist Kevin Peterson, who studies microRNAs, lamented, "I've looked at thousands of microRNA

genes, and I can't find a single example that would support the traditional tree." According to the article, microRNAs yielded "a radically different diagram for mammals: one that aligns humans more closely with elephants than with rodents." Peterson put it bluntly: "The microRNAs are totally unambiguous… they give a totally different tree from what everyone else wants."[80]

Similarly, a 2012 paper states that "phylogenetic conflict is common, and frequently the norm rather than the exception." These conflicts include stark differences between morphology-based and molecule-based trees: "Incongruence between phylogenies derived from morphological versus molecular analyses, and between trees based on different subsets of molecular sequences has become pervasive as datasets have expanded rapidly in both characters and species."[81]

Molecular biology cannot even determine a consistent tree of life for yeasts. A paper in *Nature* reports that the comparison of 1,070 genes in twenty different yeasts produced 1,070 different trees.[82]

Another argument offered in support of evolutionary common descent is the idea that cellular organelles like chloroplasts and mitochondria originated through symbiogenesis, that is, the emergence of a new life form through the evolutionary interdependence or merging of two or more species. Chloroplasts, which are essential for photosynthesis, and mitochondria, which convert oxygen into usable chemical energy in a cell, each have their own DNA distinct from the cell in which they live. In each case, a living cell is thought to have ingested either the chloroplast or mitochondria, but not to have digested it. Instead, a symbiotic relationship ensued that is essential to much of life on earth. Given that a particular facet of life is said to have originated from other life, one might extrapolate and make the argument that in general life comes from life.[83] But it is good to remember that extrapolations are not proofs, especially so when applied to whole new levels of complexity.

Biogeography

Molecular biology was not known in the mid-nineteenth century when Darwin and Wallace were doing their research. But it was widely known that islands often had their own distinctive flora and fauna. The obvious example is the Galápagos. But there are others. Hawaii is known for its many types of honeycreepers and fruit flies. Australia is known for its kangaroos and other marsupials. This could suggest that the present set of animals descended from earlier life-forms in that region, because the marsupials are variations on a form found in Australia and not in too many other places, just as the various genera and species of honeycreeper finches in Hawaii are understood to be variations on a primordial honeycreeper.

On the other hand, the evidence from biogeography is not as clear as Darwinists would lead one to believe. The honeycreepers, for example, all remain finches, and a particular sort of finches, honeycreepers. Even granting that they are all derived from a single parent population of finches, the amount of evolution here is interesting but falls wells short of the dramatic evolutionary innovations that would be required of universal common descent.

From here the case goes from bad to worse. South American monkeys, called platyrrhines, are thought to have evolved from African catarrhine monkeys, based on morphological and molecular comparisons. Fossil evidence of monkeys in South America dates to 30 Mya; but South America was an island continent from about 80 to 3.5 Mya, and had split off from Africa starting around 180 Mya. So the question is how the catarrhines could have arrived in South America to give rise to the platyrrhines. The current best explanation from the Darwinists is that the monkeys came over on a natural raft. But the explanation is not very convincing, because even if the distance across the South Atlantic Ocean was much smaller 30 Mya than it is now, it was still significantly more than 1,000 kilometers; and one doubts that the monkeys would have prepared for the journey by taking along jugs of fresh water, dried bananas, and other provisions. Moreover, the monkeys are not an exceptional case.[84]

Just Plain Wrong

While there is something to be said for some common-descent arguments, we need to put two arguments for common descent to rest:

Haeckel's biogenetic law and vestigial organs. Haeckel's biogenetic law (also called recapitulation theory or embryological parallelism) was inspired by Darwin's citing of embryology as a powerful support for his theory of evolution.[85] Darwin had relied on the earlier work of Ernst von Baer, an embryologist, who noted that he found it hard to distinguish the embryos in their earliest stages of lizards, small birds, and mammals. It turns out that von Baer objected to Darwin's theory and even to Darwin's understanding of what von Baer had actually meant, but that need not detain us here. For Darwin, embryology provided "by far the strongest single class of facts" to support his theory of natural selection.[86]

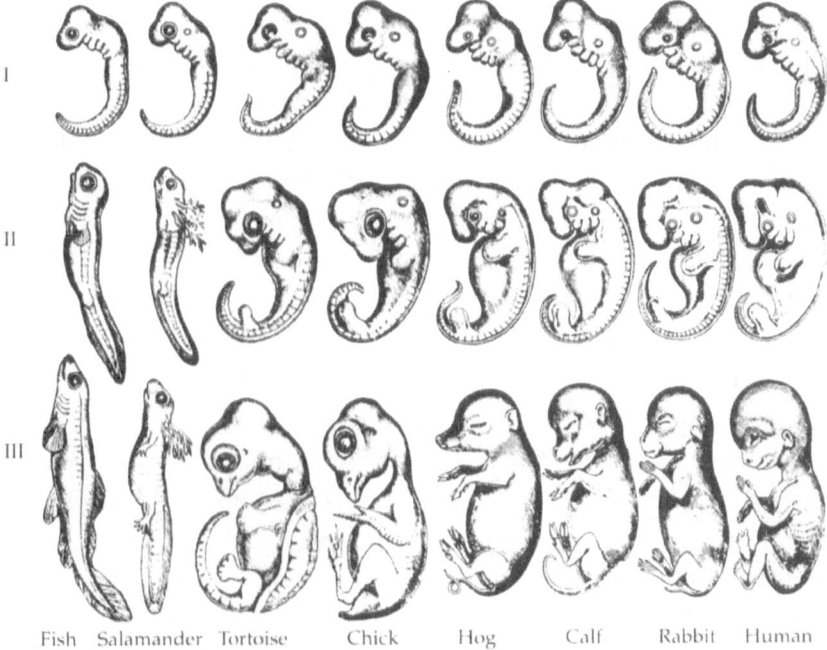

Figure 2.4.
Haeckel's embryos, as reproduced by G. J. Romanes, copied from Haeckel's original embryo drawings.

Ernst Haeckel, Darwin's champion in Germany, lent his considerable linguistic and artistic talents to the supposed connection between embryology and evolution. He coined two new terms: "ontogeny," the development from fertilized egg to maturity of a living individual, and "phylogeny," the coming to be in evolutionary history of the phylum to which the individual belongs. With these neologisms, his biogenetic law can be expressed as "ontogeny recapitulates phylogeny." By watching an embryo develop, one could watch the evolutionary stages leading to the species.

There is, however, little basis for this assertion, other than a metaphysical attachment to evolution. What of the drawings Haeckel produced in support of his argument? (Figure 2.4) The top row depicts what he called the first stage of development. All the embryos are just about the same. But that's only in Haeckel's drawings, not in reality. Everyone now agrees that there is a significant difference in the actual appearance of the embryos at that stage, although there is still some controversy as to whether Haeckel was consciously drawing fraudulent pictures or was blinded by his commitment to the biogenetic law. In any case, the actual embryos in Haeckel's "first" stage are really embryos at some middle point of their development. Figure 2.5 shows the hourglass effect. The starting points are widely different. There is some similarity in the midpoint before we get to the variety of the adult form. It is clear that the biogenetic law looks much less impressive in the light of real embryology.

Although Haeckel's drawings are now widely admitted to be misleading, somehow the concept of recapitulation has proven an irresistible temptation to most publicists of Darwinism. Biology textbooks have continued to reproduce Haeckel's embryo drawings (or lightly modified versions) as evidence of evolution long after mainstream evolutionary biologists conceded that the original drawings are wildly misleading.[87] Darwinian publicists also will point out some folds around the neck of human embryos and often refer to them as nascent "gills," as proof that we descended from fish. But the folds are not close to being gills at that stage of development. In fish, they do eventually become gills, but clearly not so in humans. So why are they included in science textbooks?

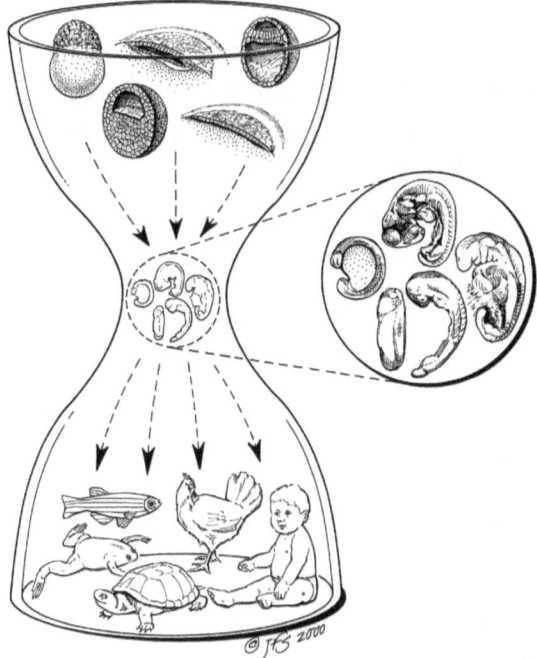

Figure 2.5.
An illustration of the fact that, contra Haeckel, embryos of various species differ markedly from each other in their early stages. (Image by Jody F. Sjogren, 2000. Used with permission.)

Miller was one of the textbook writers who featured Haeckel's drawing and presented recapitulation as an argument for evolution. To his credit, when the problems with Haeckel's drawings were pointed out to him, he substituted modern photographs and stopped promoting recapitulation. But other Darwinists, rather than admit that Miller had been teaching an erroneous doctrine, argued that the material was merely an attempt to give a historical perspective. The argument rings hollow, however, since recapitulation was presented under the heading "Data Supporting the Fact of Evolutionary Change" and the students were never told that the doctrine was erroneous and no longer believed.[88]

Vestigial organs are also (wrongly) used as an argument for evolution and common descent. This argument goes back to *The Origin of Species*, where Darwin spoke of "rudimentary, atrophied, or aborted" organs. He mentioned, as examples, nipples in male mammals, a small pelvis or hind limbs in a snake, and teeth in fetal whales. He interpreted these organs as holdovers from ancestral forms, organs that served purposes that are no longer needed. Cornelius Hunter exposed the problem with this logic in *Darwin's God*. He notes that in 1895, Ernst Wiedersheim published a list of eighty-six organs in the human body that were thought to be vestigial, such as the pineal gland, the coccyx, and the thyroid gland. In 1981, S. R. Scadding, a zoologist, went over Wiedersheim's list and concluded that just about all of them served some purpose.[89] The "vestigial" organs turned out to be not instances of useless evolutionary leftovers but rather a measure of ignorance and a readiness among biologists to fill that ignorance with what we might call Darwinism-of-the-gaps.

Evolution: Theory or Paradigm?

In the epigram to this chapter, Giuseppe Sermonti opined that John Paul II was misinformed about the theory of evolution. According to Sermonti, there is no theory of evolution:

> Evolutionism is really more of a paradigm or methodology than a theory. For its present-day supporters, the important thing about evolution is that it was due to natural causes. That the process has been continuous, unremitting, and gradual, and improving organisms all the time, is of little interest to the supporters of the theory. Some even doubt whether such characteristics are real, while others consider them as irrelevant.
>
> As a matter of fact, all the currently envisioned physical causes of evolution are either degenerative or conservative; therefore not one of them guarantees passage from the simple to the complex, from the inferior to the better. Among them there is a vague promise of progress, but in a tautological and clumsy sense; and there is concession to gradualism, though empirical observation seems to exclude this, or at least fails to demonstrate it.[90]

Sermonti is especially skeptical of the focus on molecular biology. Precisely because he is well-versed in it, he insists that DNA cannot answer the mystery, "Why a fly is not a horse." Both share many of the same genes. But there is no gene for flyness or horseness. The morphological origin of the various species of animals and plants continues to be a mystery.

In the next chapter, we will look more closely at the theory of intelligent design. Here, it suffices to underscore that there are non-ID scientists who want to avoid the Mind behind the universe, who nevertheless are getting very skeptical of Darwinism. The interested reader will find many details in Suzan Mazur's *The Altenberg 16: An Exposé of the Evolution Industry*. The book takes its title from a meeting of sixteen scientists in Altenberg, Austria in 2008, who were all convinced that neo-Darwinism was exhausted and new ideas were needed to explain the origin of biological form. Massimo Piattelli-Palmarini, who was not at the conference, revealed to Mazur why, despite the lack of evidence for Darwinism, even those who know better still defend it in public: "I think that abandoning Darwinism (or explicitly relegating it where it belongs, in the refinement and tuning of existing forms) sounds anti-scientific. They fear that the tenants of intelligent design and the creationists (people I hate as much as they do) will rejoice and quote them as being on their side. They really fear that, so they are prudent, some in good faith, some for calculated fear of being cast out of the scientific community."[91]

Keep all these things in mind the next time you hear a Darwinist confidently assert that "evolution has been responsible for X, Y, or Z."

The Origin of Life

The *Origin of Species* was about changes in existing life-forms. Natural selection could only select those plants and animals that had some present breeding advantage. But to breed, one has to be alive. So it is not surprising that Pierre Larousse's *Grand Dictionnaire Universel du XIXe Siècle* declared that spontaneous generation is a "philosophical necessity," whatever Louis Pasteur's evidence to the contrary might indicate. Or to speak in more contemporary language, some purely materialistic process for the origin of the first life is an essential step for the materialist on the way to the theory of biological evolution.[92]

In *The Origin of Species* Darwin limited himself to only one brief passage about how the first life-form might have arisen. The very last sentence of the book reads: "There is grandeur in this view of life, with its several powers, having been originally breathed into a few forms or into one; and that, whilst this planet has gone cycling on according to the fixed law of gravity, from so simple a beginning endless forms most beautiful and wonderful have been, and are being, evolved." In the second edition, as a sop to religious believers, Darwin added that the "breathing" had been done "by the Creator." But in his private correspondence, he offered a wholly materialistic explanation. In a February 1871 letter to J. D. Hooker, Darwin addressed the objection that if a materialistic origin of life from non-life were feasible, surely we would see such a thing transpiring in our day, and yet we do not. The difference, he surmised, was that in our day any such nascent new proto-life-from-non-life would immediately be met by and destroyed by an environment teeming with more evolved life-forms, giving the nascent proto-life no chance to progress further. But how would things have stood in the distant past, before the very first organism had emerged? Here is how Darwin framed the matter to Hooker:

> It is often said that all the conditions for the first production of a living organism are now present, which could ever have been present. —But if (& oh what a big if) we could conceive in some warm little pond with all sorts of ammonia & phosphoric salts,—light, heat, electricity &c present, that a protein compound was chemically formed, ready to undergo still more complex changes, at the present day such matter w^d be instantly devoured, or absorbed, which would not have been the case before living creatures were formed.[93]

Today, taking their cue from Darwin, various textbooks and museums give the impression that we know how life arose in some warm pond with methane, hydrogen, carbon dioxide, and perhaps a dash of lightning. But the truth is that there remains extraordinarily little empirical evidence for the purely materialistic origin of life and a growing body of empirical evidence that counts strongly against the idea.

Abiogenesis

Abiogenesis is the theory that living beings can arise from non-life. Perhaps a more recognizable expression for it is "spontaneous generation," although this latter term tends to refer to something that was thought to be a common natural process, rather than a one-off kick-start to evolution that moderns mean when they speak of abiogenesis. Aristotle is usually credited with the theory of spontaneous generation, with his observations that sometimes fish appear in puddles and that scallops appear in sand. The theory met no serious questioning until the seventeenth century. Maggots were thought to be spontaneously generated in rotting meat; and, mice, in barns containing wheat. Saint Thomas Aquinas accepted spontaneous generation as a fact. In the *Summa Theologiae* he wrote that animals produced through putrefaction are not produced from seeds, but through the influence of heavenly bodies.[94]

In 1668, Francesco Redi predicted that no maggots would form in rotting meat, if flies were prevented from getting at it. He put meat in three jars. He sealed one with a cork; one with gauze (letting air through); and one he left open for flies to get through. His prediction proved true: no maggots without flies. Others tried similar experiments with variable results. The usual procedure was to sterilize a solution such as broth by boiling it and then to see whether any life—usually some sort of mold—would arise from it. Life often appeared in these experiments, either because the broth was not boiled for sufficiently long or at high enough temperatures or because microbes in the air would seed the sterilized mixture. In 1859, Louis Pasteur, using a swan-necked test tube, which kept ambient air out of the sterilized broth, managed to convince most people that life could only arise from life.

Actually, the question was not settled so quickly, because some others could not reproduce Pasteur's experiment. Eventually, in 1877, John Tyndall found that microbes associated with hay could not be killed by simple boiling. But once the broth was put through a series of boil and cool cycles, no life appeared; and all were content to agree that life can arise only from life.[95] If Darwin was hoping for clear evidence that it was no problem for primordial life to arise, unaided, from a lifeless pond, then this discovery was, well, a problem.

In 1868, before Tyndall's experiments put the nail in the spontaneous generation coffin, but well after Pasteur's publicized 1859 results, the situation started to look brighter for the scenario that Darwin would suggest to Hooker in 1871. In 1868 Thomas Huxley analyzed some sediments, not from a warm pond, but dredged up from the bottom of the North Atlantic. He found some "granule heaps" surrounded by viscous matter which he described as "lumps of a transparent gelatinous substance" and a "colourless, structureless matrix."[96] He thought that he was dealing with a primitive protoplasm, the link between inorganic matter and life. He even gave this new organism a scientific name—*Bathybius haeckelii*—life from the depths, in honor of Ernst Haeckel, a fellow materialist crusader and Darwin's German publicist.

Several years later, in 1875, much to his embarrassment, Huxley had to admit that *Bathybius haeckelii* was nothing but precipitate formed by the mixture of seawater and alcohol used to preserve the samples dredged from the depths. There was no arguing with the chemists. Huxley readily admitted his mistake. Haeckel, in contrast, held on to *Bathybius* as an important element of his materialist philosophy for another ten years.

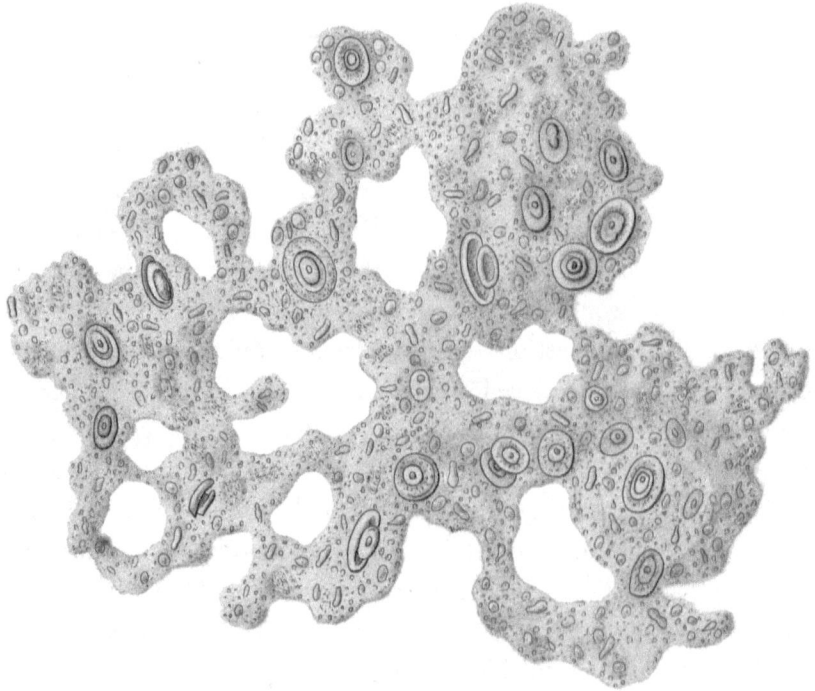

Figure 2.6.

Bathybius haeckelii, the (incorrectly) supposed bridge between inanimate matter and life.

Philip Rehbock, in his analysis of this episode of scientific "discovery," cites Loren Eiseley's characterization of the affair as "one of the most peculiar and fantastic errors ever committed in the name of science," and his explanation that it was "the product of an overconfident materialism, a vainglorious assumption that the secrets of life were about to be revealed." Rehbock does not dispute the analysis but merely states that a part of the problem was that the experts in biology did not take into account the potential contributions of the chemists before publicizing their "great" discovery.[97] A failure to consult researchers in other fields with relevant expertise is a common mistake in the field of abiogenesis, as we will soon see.

The Darwinists next turned to thought experiments to explain how the first life-form might have emerged. Alexander Oparin, in the Soviet Union in 1924, and John Haldane, independently in England in 1929, proposed that life could have originated on the early earth provided there was a reducing (oxygen-poor) atmosphere with lots of carbon, hydrogen, nitrogen, and other elements necessary for life. If, in addition to these elements, there was lots of energy in the form of lightning, ultraviolet radiation, and a high ambient temperature, then many organic compounds might have been synthesized, which eventually organized themselves into something that natural selection could work on. Haldane coined the term "prebiotic soup" to describe these conditions. The Oparin-Haldane theory became popular with hard-core materialists. My mother's skepticism toward it was one of the reasons she was barred from medical school for a few years in Communist Czechoslovakia in the early 1950s.

In 1952, Stanley Miller, a doctoral candidate, working at the University of Chicago under the Nobel laureate Harold Urey, performed an experiment to see if something like Haldane's prebiotic soup could be coaxed into producing some organic compounds necessary for life. He put methane, ammonia, hydrogen, and water into a closed system, which he heated so as to produce a gas, and then zapped it with electric sparks. The gases were then allowed to condense, only to be heated and run through the system again. After eighteen days, he analyzed the various compounds in the system. Although he could not identify all the synthesized products, it was clear that there were many different kinds of organic polymers, some fatty acids, hydroxyl acids, amide products, and—most notably—several amino acids. This was thought to be a great step on the way to confirming the prebiotic soup hypothesis. It inspired many chemists to get in on the act, conducting similar experiment with various refinements and working out how these molecules might be assembled into simple self-replicators so that natural selection could begin to work on them.

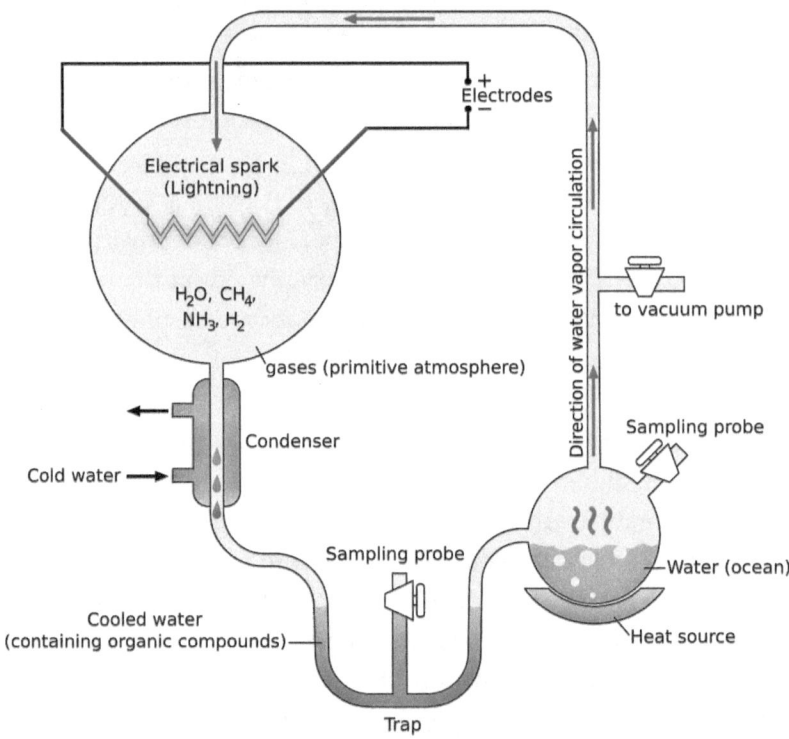

Figure 2.7.
Schematic of the Miller-Urey experiment.

In 1969, Dean Kenyon published *Biochemical Predestination*, which became a popular textbook arguing for abiogenesis. But he began having second thoughts, and by the mid-1970s, he had completely changed his mind. In 1984, he wrote the foreword to *The Mystery of Life's Origin*, which critiqued the standard materialist story. There he listed several strong reasons for his skepticism.

First, there is the problem of potentially interfering cross-reactions. So what if a few amino acids are formed? How do you get them to join into a polypeptide chain in the presence of other chemicals who want to dissolve them?

Second, the most complicated structures synthesized under pristine conditions according to strict protocols in the lab are orders of magnitude simpler than the simplest real cells.

Third, there is the problem that synthetically formed chemical compounds are racemic mixtures, which means that a near-equal number of "right" and "left" versions of the same chemical compound are formed. (D-glucose and L-glucose, for example, have the same chemical formula but not the same spatial organization, just like a human right hand and a human left hand have all the same functions but are mirror images of each other.) But life seems to require that the essential biological molecules are chiral (from the Greek word meaning "hand"). Proteins are formed from L-type amino acids whereas nucleic acids are formed from D-type sugars. They need to be "handed" in this way for the cell to work, and yet there is no known natural process that produces them thus, outside of living systems.

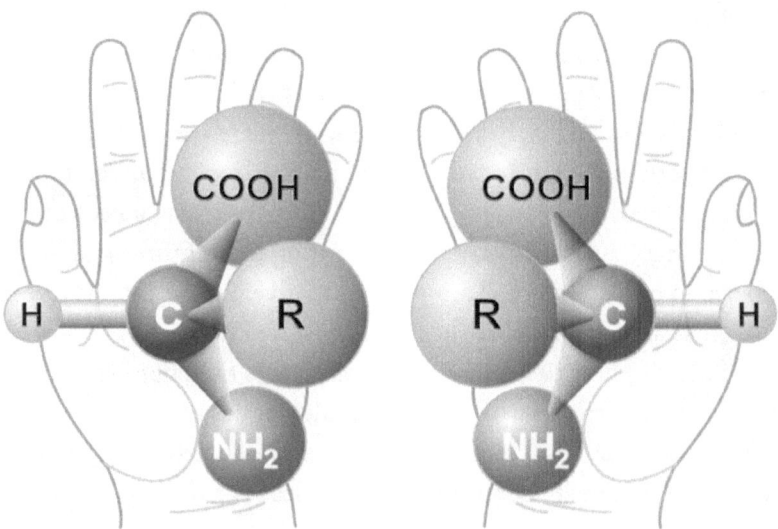

Figure 2.8.
Amino acid chirality.

Finally there is the problem of the origin of genetic information in biopolymers. As Kenyon noted, "No experimental system yet devised has provided the slightest clue as to how biologically meaningful sequences of subunits might have originated in prebiotic nucleotides or polypeptides."[98]

The *Mystery of Life's Origin* was republished in 2020, with a new section that takes into account what has been learned since 1984. In a chapter entitled "We're Still Clueless about the Origin of Life," the internationally renowned synthetic organic chemist James Tour writes:

> Two-thirds of a century since the 1952 Miller-Urey experiment, where some racemic amino acids were formed from small molecules and an electrical discharge, the world is no closer to generating life from small molecules—or any molecules for that matter—than it was in 1952. One could argue that origin-of-life research is even more befuddled now than it was in 1952 since more questions have evolved than answers, and the voluminous new data regarding the complexity within a cell makes the target much more daunting than it used to be.[99]

Tour appeals to the origin-of-life (OOL) research community to "step back and consider the claims within the research, the true state of the field, the retarded state of the science relative to other research areas, and the confusion or delusion of the public regarding life's origin." He acknowledges that many origin-of-life researchers are superb scientists, but adds that "overly confident assertions, exaggerated and spread by the overzealous press, have led to gross misconceptions regarding what is and is not known concerning the beginning of life."[100] To back up his statement, Tour goes on to give many examples which the technically inclined will want to read for themselves. As in the case of Huxley and the *Bathybius haeckelii*, it is the chemists who are blowing the whistle on the materialist ideologues.

In August 2023, Tour increased the pressure on origin-of-life scientists to acknowledge the basic difficulties confronting abiogenesis. He issued a personal challenge to ten top researchers in the field to explain how just one of five essential chemical processes could have arisen through an unguided natural process. He offered to take down all his internet posts about OOL research and evermore remain silent on the subject if any three of these top ten chemists agreed that a proposed explanation was correct within sixty days of his challenge.[101] Not one of them rose to the challenge.

Fred Hoyle was a non-religious scientist who admitted that life is just too complicated to have arisen spontaneously from basic chemistry. He thought it more likely that a tornado ripping through a junkyard would assemble a Boeing 747 airplane than that a cell would be assembled by random by random assemblages of molecules in a prebiotic soup. Francis Crick thought that perhaps life originated in some friendlier place in the universe and the subsequently evolved creatures sent a capsule with some microbes in it toward Earth, a planet they identified as being able to sustain life. The space capsule duly arrived with its precious cargo, and natural selection took care of the rest to produce the variety of flora and fauna that has ever inhabited our planet. This scenario is called "Directed Panspermia." One knows that origin-of-life research is nowhere when a Nobel laureate has to resort to such nonsense to argue for Darwinian materialism.

Even the prestigious journal *Nature* recently admitted to the impasse in origin-of-life studies: "Water is essential for life, but it breaks down DNA and other key molecules. So how did the first cells deal with such a necessary and dangerous substance?"[102] The short answer is that no one knows, because if they did, they would surely have given us an answer.

A still more recent article in *Nature*, by Nick Lane and Joana Xavier, also acknowledges the problems. "Explaining isolated steps on the road from simple chemicals to complex living organisms is not enough," they wrote. "Looking at the big picture could help to bridge rifts in this fractured research field."[103] More specifically, as David Coppedge notes, "They point out that no one has figured out (i) the target of selection, (ii) what to look for, or where, (iii) whether microfossils are biological or not, (iv) where genes and proteins came from, (v) how to unify the splintered field of OOL research, and

(vi) how to build a coherent framework for OOL." They also point out the major problems with the most popular approaches. Coppedge summarizes them thus:
- prebiotic soup model: implausible sources of ingredients, concentrations, and longevity.
- RNA world: ribozymes tend to disintegrate and randomize, not grow in complexity.
- hydrothermal vents: no plausible origin of metabolism or polymerization.[104]

My favorite bit of logic meltdown on the subject of abiogenesis comes from Mark Ridley, a well-known evolutionist. After admitting (in 1985) that we have no idea how the first cells originated, he added, "For the pure nihilist, this is all very well. But for anyone who wants to understand the origin of life, it is something of a paradox. Natural selection is the only theory that we have to explain evolution; an explanation, if it is going to come from anywhere, will come from natural selection."[105] Why should natural selection, which selects the fittest replicator, be the only possible cause of assembling a replicator in the first place? Far from the only possible cause, it is a decidedly inadequate cause. Natural selection can only work on something that replicates because it is all about differential rates of reproduction. Why would a process that can only work on replication produce a replicator in the first place? Ridley is blinded by his commitment to natural selection. Could there be a clearer warning of what dogmatic Darwinism can do to one's critical faculties?

Summary

As John Paul II acknowledged, there is some evidence that some if not all of today's life-forms are descended from earlier living forms. That would be "evolution" in the more general sense. Darwin's proposed explanatory mechanism is also a well-founded inference within limits. But it is very far from the whole story. And it is by no means certain that all life-forms descended from one original biological self-replicator.

Not everyone will agree with this assessment. It will not convince the likes of Dawkins and the vociferous atheists. Fodor and Piattelli-Palmarini were criticized for bashing a straw man: the outcry was that no one is *that* kind of Darwinist anymore. But they had no trouble presenting lots of evidence that there are still *those* kinds of Darwinists in positions that shape public opinion. And even if such people have some hesitation about the full scope of Darwinism, they instinctively recite the materialist *credo*—defend Darwin—if they think its critics are trying to challenge a purely material explanation for everything. Fortunately, if even the *Economist* admits that biologists are now starting to speak heresy by positing alternative explanations for the evolution of cichlids and other animals, there is some hope that Darwinism's stranglehold on the culture will eventually be broken.

3. Intelligent Design in Nature

The only satisfactory explanation of the origin of such "end-directed, self-replicating" life as we see on earth is an infinitely intelligent Mind.[1]
—Antony Flew, one of the world's most notorious [former] atheists

In the last chapter, we saw that science knows next to nothing about how purely natural forces could originate the first cell and that, at best, natural selection can only explain limited variations in plant and animal life. Contrary to what Mark Ridley asserted at the end of the previous chapter, there is an alternative to a choice between a nihilistic shrug and the dogmatic assertion that natural selection, despite a glaring lack of empirical evidence, somehow explains the origin of life. But that alternative does require an openness to go beyond scientific or methodological naturalism. The alternative can broadly be called intelligent design (ID), insofar as it allows for the creative involvement of an intelligence, a mind. But there are different versions of ID, so it is necessary to make some distinctions.

Some Basic Philosophy of Science

Before we can speak of ID, it is important to know something of the philosophy of science. Most people will find it surprising that philosophers cannot provide a rigorous definition of science, despite its importance in modern society. The various technologies that it spawns are a clear indication that science has a real grip on the world around us. Science is seen as the very paragon of rationality and by many as the ultimate authority in today's world. But what is science? What do physics and gender theory have in common? And how is it that the "settled science" of yesterday is often dismissed as nonsense today? These are questions that anyone trying to understand science will need to address. Fortunately, for our purposes, we do not need to answer all of them. Rather we can focus on just a few characteristics of science as applied to the material world in which we live.

The word "science" comes from the Latin word *scire* meaning "to know." The medieval Scholastic philosophers, following in the footsteps of Aristotle, distinguished *scientia* from belief or opinion. On this view, science was reserved for knowledge that could be demonstrated: knowledge through causes. Mathematics was also widely regarded as an example of science, even though it rarely invokes causes of any kind. The medievals also used science to describe various branches of theology. True knowledge was believed to be attainable in theology because the premises came from God's revelation and the rules of logic. Although the medievals thought that one could attain demonstrative knowledge in cosmology, they called those who spent their time pondering local motion or plants and animals "natural philosophers" and not "scientists." The term "scientist" was coined in the 1830s, and in the course of the nineteenth century the modern meaning of science as a systematic and empirical study of nature coalesced.

The very first controversy about the modern enterprise called science concerns its roots or origins. There is no doubt that the seventeenth century saw a great acceleration in the development of mathematics and physics. At the beginning of the century, the Ptolemaic system was still dominant, the kinematics of falling bodies and projectile motion could not be described by mathematical equations, and the mathematical tools themselves were not up to the job. René Descartes had yet to translate geometry into algebra, and calculus had yet to be developed. The microscope had just been invented but not the telescope. Missing too were ideas about light that would permit anything but the most basic calculations and predictions.

By the end of the seventeenth century, the sun was understood to be at the center of the solar system, with the planets orbiting around it in elliptical paths, as described by Newton's equations of motion. The planets were following the same laws of motion as apples dropping from trees or cannonballs shot in the heat of war. Newton and Leibniz had independently developed calculus. The telescope had revealed mountains and craters on the moon and moons circling other planets. And white light had been decomposed into the spectrum, which was then recombined to produce white light. No doubt, these were major accomplishments.

Many historians have celebrated the seventeenth century as the great scientific revolution. But did science "come down from heaven to earth along the inclined plane of Galileo," as Henri Bergson would have it?[2] Or was it a flowering of seeds laid down earlier, its roots stretching deep into the Christian Middle Ages? Pierre Duhem, a pioneer philosopher and historian of science, argued for continuity rather than revolution. But some prominent historians of science of the twentieth century, notably George Sarton, tried their best to keep his work from being published, because it contradicted their picture of Christianity as a hindrance to science. We do not need to be distracted by this controversy. Suffice it to say that the brightest stars of seventeenth-century science were, for the most part, serious Christians. This is something that the atheists of today seldom mention or can even begin to appreciate.

Physics was the first of the sciences to be developed, although the seventeenth century saw important progress in other fields as well. William Harvey, for example, correctly described the circulation of blood, and Francesco Redi, as we saw in the previous chapter, showed that maggots were not spontaneously generated from rotting meat. Nevertheless, physics led the way; and the crowning achievement of this enterprise was Newton's *Principia Mathematica Philosophiae Naturalis*.

The *Principia* contains a short section entitled "Rules of Reasoning in Philosophy." Two rules are especially relevant for our purposes. The third rule states, "The qualities of bodies, which admit neither intensification nor remission of degrees, and which are found to belong to all bodies within the reach of our experiments, are to be esteemed the universal qualities of all bodies whatsoever."[3] What this means, according to Newton, is that "the extension, hardness, impenetrability, mobility, and inertia of the whole, result from the extension, hardness, impenetrability, mobility, and inertia of the parts." This is the list of the "primary" qualities, which supposedly had the power to induce sensations of the "secondary" qualities in us, qualities such as colors or taste or smell. Newton may not have seen the full implications of this way of looking at things, but he provided later physicists a framework and mathematical tools to explore the material world. Newton's third rule was a powerful approach to solving equations of physics. One just had to add up—to integrate—the contributions of each of the parts to describe the behavior of the whole. The calculations were still difficult—often intractable—but in theory the whole world could be described by the resultant equations.

In this model of the world, as it was developed in succeeding generations, there was nothing over-and-above atomic particles interacting in a void to account for color or taste or smell—or consciousness, for that matter. But this was deemed a small price to pay for the world's being amenable to mathematical description. The natural philosophers of the era could apply the math to many different situations and it proved correct. The period of a pendulum could be calculated based on its length and mass. The prediction that one's weight on the equator is less than one's weight in England was verified. The mathematician Joseph-Louis Lagrange spoke for many when he said: "Newton was the greatest genius that ever existed, and the most fortunate, for we cannot find more than once a system of the world to establish."[4]

Newton's fourth rule for reasoning states, "In experimental philosophy we are to look upon propositions collected by general induction from phenomena as accurately [sic] or very nearly true, notwithstanding any contrary hypotheses that may be imagined, till such time as other phenomena occur, by which they may either be made more accurate, or liable to exceptions."[5] This is an important support for the famous hypothetico-deductive method, which is often presented as the hallmark of science. A natural philosopher was supposed to observe nature and come up with a hypothesis, H, to explain its functioning. Next, from the hypothesis, H, he was supposed to deduce a hitherto unobserved result, R, and then perform an experiment to test the prediction. If the experiment produced R, then he could hold on to H, and he would be more confident that H was correct.

As a piece of logic, the hypothetico-deductive method does not work, because there might be some other hypothesis, call it G, that could equally well predict the same R. Moreover, the method suffers from the problem of induction, which from a few test cases—even many but always a finite number—jumps to a universal conclusion. Newton was aware of this problem but recognized that if the natural philosopher wanted to make progress, he must not allow himself to be distracted by the quibbles of logicians.

Although the hypothetico-deductive method could not prove a hypothesis to be correct, it was useful in rejecting a bad hypothesis. The philosopher Karl Popper latched on to this trait to distinguish between true science and pseudoscience. A true science, according to Popper, is one that is not afraid of making predictions, which, if proved wrong, would lead it to revise its hypotheses. Physics serves as the paragon of true science because its practitioners make predictions before the fact. Einstein, for example, predicted that light could be bent by gravity and that the next solar eclipse could provide an opportunity to put his theory to the test. It passed. This was different from after-the-fact rationalizations provided by various pseudosciences.[6] It should be clear from the evidence discussed in the previous chapter that Darwinism fails Popper's test. It predicts that transitional forms should be found. These have not been found after more than a century and a half of looking; and yet Darwinists insist that their theory is correct.

The hypothetico-deductive method Newton described came to be known interchangeably as Newtonian science, empirical science, and Baconian science. Francis Bacon was the author of the *New Organon*, which he wrote to argue against Aristotle's notion of natural philosophy.

He accused the Aristotelians of using words—formal causes—to disguise ignorance. In the same vein, Molière lampooned the Aristotelian doctor who pronounced that opium puts people to sleep because of its dormitive properties. (Since "dormitive" means "sleep-inducing," saying that a substance puts people to sleep because it has a sleep-inducing property is no explanation at all.) No synthetic sleeping pill could ever be developed from such a purely verbal level of understanding. Bacon was also keen to rid natural philosophy of final causes. One does not gain much specific insight by saying that the butterfly was made for the glory of God. That could be said of anything. Once formal and final causes were banished, purpose was stripped from nature and only the material and efficient causes remained. This is exactly the presupposition of scientific materialism. Of course, many scientists who followed Bacon—notably Newton himself—maintained purpose in nature, though in a form quite different from the Thomistic and Aristotelian view that preceded them. It took centuries for this banishment of purpose to conquer natural science altogether. It culminated with the work of Darwin.

In its hardened form, materialist science makes a claim about what exists, and what we can know. In short, nothing exists except what can be described by physics, and physics can never point beyond itself; so it may be called physical reductionism. I encountered this philosophy in the very first class I attended in university. It was a lecture in physics; and the professor began by congratulating us for choosing to study physics. Chemistry, he said, is really physics because all reactions follow the rules of quantum mechanics with perhaps some future tweaks necessary to account for gravity. Chemistry is physics for the mathematically challenged. But give it some time and physicists will develop an enhanced quantum mechanics that can handle gravity, and mathematicians will develop more powerful means of calculation and then chemistry will no longer be a separate department at the university. Biology will be the next discipline to be integrated, because life is chemistry, and chemistry is physics. Finally, the humanities too will be eliminated because man is just an animal, and an animal is just fancy biology, and biology is just complex chemistry, and chemistry is physics. So congratulations for choosing physics, because physics is everything.

I was taken in for a while by this propaganda, but fortunately grace, building on common sense, opened my eyes. Sadly many influential spokesmen for science believe physics can fully explain everything and, indeed, must. Perhaps the clearest avowal of this attitude is an often-quoted directive from Marxist evolutionary biologist Richard Lewontin:

> It is not that the methods and institutions of science somehow compel us to accept a material explanation of the phenomenal world, but, on the contrary, that we are forced by our *a priori* adherence to material causes to create an apparatus of investigation and a set of concepts that produce material explanations, no matter how counter-intuitive, no matter how mystifying to the uninitiated. Moreover, that materialism is absolute, for we cannot allow a Divine Foot in the door.[7]

Lewontin is not alone. This is the attitude of mainstream academia and, as we will soon see, the costs of dissension from it are often high. It accounts for why there is a tremendous opposition to introducing an intelligent agent into the explanatory framework of nature.

Intelligent Design—The Explanatory Filter

The term *intelligent design* is rather clumsy, for in ordinary parlance, design is understood to be the product of intelligence, so "intelligent design" is a bit like "full-contact wrestling" or "mind-generated art." But it was chosen because Richard Dawkins and most of academia insist that natural selection, though unintelligent and blind and unconscious, has produced many plants and animals that have the appearance of design: "designoids" is what Dawkins calls them. "Intelligent Design" was, if not coined by the biochemist Charles Thaxton in 1988, certainly introduced into common circulation by him at that time. The intelligent design (ID) community coalesced a few years later, in 1993, at a meeting in Pajaro Dunes, California. Phillip Johnson, a Berkeley law professor and author of the bestselling *Darwin on*

Trial, spearheaded a gathering of scientists and philosophers who had serious reservations about Darwinism.

There was Michael Behe, a molecular biologist from Lehigh University in Pennsylvania. As a young scientist, Behe had accepted the Darwinian account of evolution with hardly a second thought. He did not have to have second thoughts, because the theory was irrelevant to his work in molecular biology. But then, in 1985, he read Michael Denton's *Evolution: A Theory in Crisis*. He was impressed by the book and then angry that none of the problems of Darwinism had ever been mentioned to him in the course of his education. He began to think about evolutionary theory in light of his own work and eventually wrote the influential *Darwin's Black Box* (1996).

Behe was joined in Pajaro Dunes by Dean Kenyon, who had written a bestselling textbook in 1969, *Chemical Predestination*, in which he argued for a purely materialist account of the origin of life. But as noted above, by 1993, Kenyon had long since changed his mind. The next year, he would be temporarily suspended from teaching at San Francisco University for voicing his doubts about the Darwinian paradigm while teaching a class of undergraduates.[8]

The meeting was also attended by William Dembski, a mathematician and philosopher interested in the logic of detecting design who would go on to publish a monograph on the subject with Cambridge University Press. Stephen Meyer was there with a background in geophysics, a keen interest in origin-of-life chemistry, and a doctorate in the history and philosophy of science from the University of Cambridge. Present too was Jonathan Wells, with PhDs in biology and theology (from Berkeley and Yale respectively) who went on to write the bestselling book *Icons of Evolution*. All told, there were fourteen attendees, most of whom took the earth and the universe to be billions of years old,[9] though Johnson urged the group not to make the age of the earth a central issue but rather to focus on the evidential deficiencies of Darwinism and the positive evidence for intelligent design in nature.

Most of those who met at Pajaro Dunes came to be associated with the Seattle-based Discovery Institute and its Center for Science and Culture. Given the small number of people initially involved, it is amazing how much success they have had in disseminating their ideas. Three decades on, they have an excellent website that sees significant daily traffic. They hold large conferences in major cities in the United States and abroad. They have published several influential books. They produce superb videos making the case for intelligent design, most of which are available at their Discovery Science YouTube page. And they and other scientists who have joined the movement have published scores of peer-reviewed science articles that contribute to the case for intelligent design.

Some intelligent design proponents, such as Michael Behe, emphasize irreducible complexity as a telltale marker of intelligent design. Others, such as Dembski and Meyer, subsume irreducible complexity under the more general concept of specified complexity—a type of information. They note that the world contains an immense amount of such information, most famously but not exclusively in the DNA of all living things. They then argue that no unintelligent process can account for this information, while intelligent design can.

Their thought process is akin to that of a person innocent of Russian literature and its leading authors coming across a copy of *War and Peace*. The book's front matter is missing and there are no running heads. In consequence, there is nothing about an author listed. Nevertheless, the person who has found the book rightly concludes that it was surely composed by someone, because the text could not be produced by the hydrodynamics of ink on paper nor by monkeys bashing away at a keyboard. Of course, the information we encounter in living organisms (e.g., DNA) is tied up with the ability to reproduce, so this adds a layer of complexity to the matter missing from our *War and Peace* illustration. So let's take a closer look at ID, as conceived by Discovery Institute's top scholars working in this area.

The logic of the argument, as we have seen, was given a mathematical treatment by Dembski in his monograph, *The Design Inference*, first published in 1998 and then in a revised and expanded second edition in 2023. The details can be found there, but the following should suffice. The design inference is based on what Dembski calls the explanatory filter. A filter separates. And this logical filter separates the designed from the necessary or the merely fortuitous.

If we come upon something and want to know whether it is a product of intelligence or nature, we begin by asking whether there are some natural laws that account for its structure or information content. So, for example, ice crystals or the regular pulses from a quasar could be explained by known natural laws and so would not demand the direct input of an intelligent agent.

Theists, of course, insist that the laws of nature are themselves caused by an intelligent agent—God—but we also recognize that God, in many instances, works by secondary or instrumental causes. Sometimes, as in the case of physical laws, the instrumental causes are not themselves intelligent. That is, they are not mindful beings. But sometimes, as in the case of the inspired authors of Scripture, the secondary causes are intelligent. God is the primary author of Scripture because he inspires the various human authors—Moses, David, Matthew, etc.—to convey the truths He wants to reveal. These human authors are intelligent instruments used by God. The ID explanatory filter is intended to determine whether the detectible immediate cause is intelligent. I specifically chose "immediate" cause rather than "secondary" in the last sentence, because the filter does not rule one way or the other on whether an effect involved both a primary/divine cause and a secondary cause. It is only concerned with whether the effect in question demands an intelligent agent. If, in a case under consideration, natural law suffices to explain the artifact or event, then we have an unintelligent immediate cause, and the filter would stop at "necessity." (Again, this does not rule out God both causing and superintending the natural process.)

If no natural law can account for the form of the object under examination, we have contingency and we come to the next stage of the filter. Paleontologists often find themselves in this position. They find a rock with one or more lines on it. Are the lines the work of some ancient human or other hominid? Or are they the result of some chance natural process? Is the rock some primitive cutting tool, or is it a rock untouched by any intelligent animal?

Design Detection

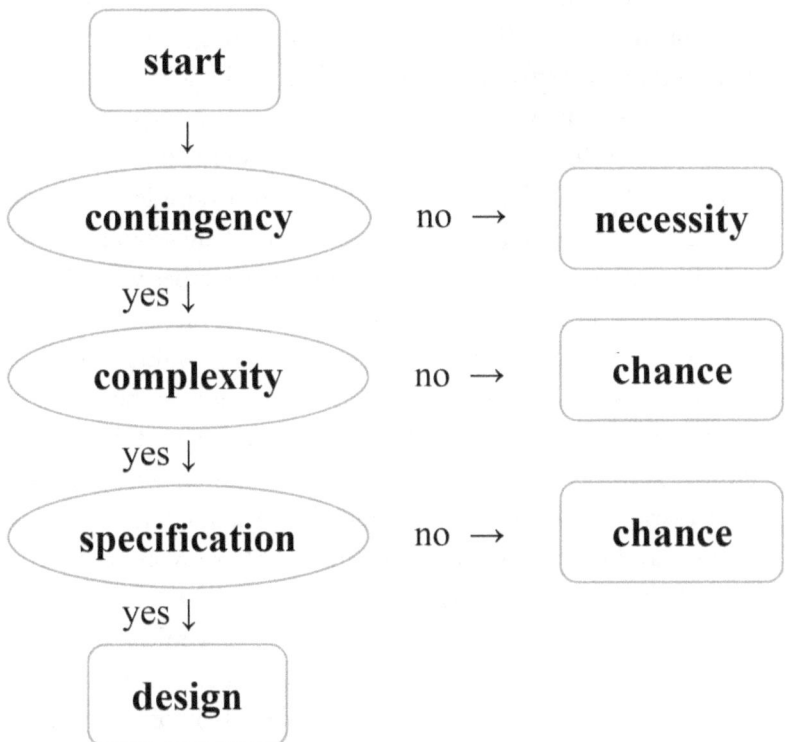

Figure 3.1.
An illustration of mathematician and philosopher William Dembski's explanatory filter. In his book *The Design Inference*, he sets the threshold for what constitutes complexity extremely high, such that the filter allows for some false negatives (i.e., identifying something as undesigned that was designed) while reliably avoiding false positives (i.e., identifying an undesigned object as designed).

The explanatory filter, at this stage, asks about the complexity of the object or process. Most paleontologists would throw away a stone with a single scratch, chalking it up to chance. (Dembski, I should note, later streamlined the explanatory filter, deftly collapsing chance and necessity into a single category. I will leave the interested reader to explore the details of it, which can be found in the second edition of *The Design Inference*, co-authored with software engineer Winston Ewert.) But if there were several scratches on the rock, the paleontologists might have a second look. The difference is complexity.

Complexity by itself, however, is not enough. The clinching evidence—the last step of the filter—is whether the scratches manifest specified complexity. A specified pattern matches an existing pattern, such as a meaningful message. If the several scratches formed a known Egyptian hieroglyph, here we would have an example of specified complexity. A less obvious case, but one that likely would still trigger a design inference, would be if the several scratches were lined up perfectly at right angles so as to produce three crosses spaced equally apart across the face of the rock. The paleontologist would probably discard the stone with the random scratches, viewing it as the product of chance, whereas he would no doubt publish a paper about the one with the crosses as evidence of early human tools—design. Why? Because it is both complex and specified, the telltale mark of intelligent design.

The usual illustration of the explanatory filter involves the letters of the alphabet and the phrases they compose. If we fed the typed pattern "be" into the explanatory filter, the filter would relegate it to "chance," not because the filter proved unequivocally that it was chance but because it might be due to chance and the explanatory filter is more concerned with avoiding false inferences to design than false inferences to chance or necessity. When the design filter returns a design inference, we want to be confident that it was in fact design, even if that means employing a filter that occasionally misses detecting design in a pattern that was designed. So, why would the filter fail to detect design for the letter combination "be," an English word? If we forget about capitalization but include spaces, the probability of producing the pattern "be" is 1 chance in 27 x 27, or one chance in 729, which may seem fairly improbable, but it is a lot more probable than winning the state lottery, which someone manages to do just about every week.

If we turn to Dawkins's illustrative phrase "methinks it is like a weasel," the odds of randomly typing it are, as noted earlier, one in 27^{28}, or roughly one in 10^{40}. If a monkey could type a random string of 28 letters per second, and had unlimited patience and endurance, then it would take the monkey roughly 10^{32} years to produce the phrase, which for all practical purposes means the task is impossible. (Actually, this is based on independent 28-letter attempts. The probability of finding the string in any random string of 28 letters and spaces in the growing series of letters and spaces is somewhat smaller, but not significantly so.) Note, however, that the probability of a monkey typing any 28-letter-long phrase is exactly the same. The random sequence "aidnne par aplroa dzikpladkf" would be just as difficult for the monkey to produce as "methinks it is like a weasel." The random sequence has the same complexity as the phrase from Shakespeare, but the explanatory filter would relegate it to the realm of chance because its complexity is not specified. It matches no preexisting pattern, much less a meaningful one. It makes no sense.

Something more needs to be said about information content and what it means for that information to be specified. What follows is admittedly much simplified, but this will not bias the explanatory filter to find design where none exists. On the contrary, it might hide design; that is to say, it might get the filter to relegate something that is designed into the realm of chance.

Let us begin by looking at how one might quantify information. There is an adage that "a picture is worth a thousand words," so let's take a look at the information in a digital photograph. Digital cameras are usually rated by the number of megapixels in the optical sensor. The higher the number, the bigger the print that we can make from the digitally stored image without the image beginning to look pixelated. Each pixel can be thought of as a unit of information. So the raw pixel count on the sensor is a first approximation to the amount of information the photograph contains.

Imagine taking a picture of a uniformly blue wall under uniform white light. Every single pixel in the RGB raw file is going to read (0,0,255). So even if the picture were taken with a 16 megapixel optical sensor, it could be stored accurately with just a few bytes of information. If one looks at the file sizes of pictures taken with a 16 megapixel camera, they vary in size but are usually much smaller than the 16 megapixels. I just took three such pictures on the finest setting on my camera and the resulting file sizes were between 4.5 and 7 megabytes. The reason for the shrinkage is that the files are saved in JPEG format, which compresses the information so as to make it possible to store more pictures on the SD card and cut down on data usage when sending it from one phone to another. The compressed files, however, usually lack some of the information contained in the original. Several years ago, I came across a news story entitled "JPEG is dangerous to your health." In the early days of digital X-rays, it turned out that the compression software made hairline fractures invisible.

This was a visual example of quantifying information, but we could have looked at sound recordings and compared WAV and MP3 files. Ultimately, all sensory information can be reduced to byte count. And, although I am not going to try to quantify the information content of Mona Lisa's smile, some sense of the amount of information for any given object or process can be gleaned from the total number of bits needed to store it accurately in digital form.

Now we need to look at what "specified complexity" means. A system with specified complexity needs to be exactly or nearly exactly what it is for it to function. Every English sentence, for example, has specified complexity. "Methinks it is like a weasel" has specified complexity whereas "aidnne par aplroa dzikpladkf" gives every indication to an English speaker that it is random gibberish, and, hence, unspecified.

Outside of the easy-to-understand examples of language, we can point to almost any mechanical or electrical device as an example of specified complexity. A mousetrap, to use Michael Behe's famous example, is simple relative to most machines, but it too works only if the parts are shaped and connected just right. And even the materials themselves must be just so. A mousetrap spring that was shaped precisely as it needed to be and connected just so still wouldn't work if it were made of soft rubber or some wholly inflexible element or compound. The point is that there are layers of specificity necessary for the mousetrap to work—specified complexity.

A photograph of a platform, a spring, a hammer, a catch, and a holding bar lying disassembled on a table will have about the same number of bytes as a photograph of a properly connected mousetrap.

But only if the five elements are properly connected will there be a device that has any chance of catching mice.

The five-element mousetrap is an example not only of specified complexity but also of what Michael Behe calls "irreducible complexity." As we saw in Chapter 2, he means that one cannot remove any one of the five parts and still have a functioning mousetrap. We might be able to enhance the mousetrap by adding further elements: a second spring or a more elaborate catch. But the trap will not function with any of the five parts removed. As one might imagine, there are all sorts of attempts on the web to prove Behe wrong. For example, it has been pointed out that a glue trap has fewer than five parts. But, as we will see, such objections miss the point, and no one has effectively rebutted the challenge irreducible complexity poses for evolutionary theory.

It is possible to get a sense of complexity without resorting to mathematics. Most people, for example, have an intuitive sense that typing monkeys will not readily produce "Methinks it is like a weasel," even if the average person doesn't know how to calculate the odds. Nevertheless, in a technical discussion of the power of evolution, some mathematical notions will need to be introduced.

The vast time since the first life appeared on the earth—some 3.8 billion years—and the sheer number of bacteria in an average human body—about thirty-nine trillion—make it clear that the total number of organisms that ever lived on earth is very high. A rough calculation puts the total at about 10^{40} (1 followed by 40 zeroes), mainly bacteria.[10] That's a lot of life-forms experiencing random mutations—a lot of tries at winning an evolutionary lottery by lucking into a beneficial random mutation. So it may seem that with so many organisms around, one might beat some fairly high odds stacked against Darwinian evolution.

The ID proponents are, of course, aware of all this. That is why Dembski requires that before his explanatory filter signals design, the odds against an event must be smaller than 1 chance in 10^{150}. If you are unfamiliar with how exponents work, you might imagine that this number is just under four times larger than 10^{40}. But every time the exponent (the superscripted number) increases by one, the value of the total increases tenfold. So 10^{41} is ten times more than 10^{40}, and

10^{42} is 10 x 10, or 100, times more than 10^{40}. Thus, 10^{150} is 10^{40} times a hundred, times a trillion, times a trillion, times a trillion, times a trillion, times a trillion, times a trillion, times a trillion, times a trillion, times a trillion. Dembski, in other words, has set the threshold for detecting design absurdly high, all in the interest of avoiding false positives—that is, avoiding ever labeling something the product of intelligent design where no intelligent design was involved. He calls this number the universal probability bound.

Dembski and Jonathan Witt explain the calculation of the threshold as follows:

> Scientists have learned that within the known physical universe there are about 10^{80} elementary particles.... Scientists also have learned that a change from one state of matter to another can't happen faster than what physicists call the Planck time.... The Planck time is 1 second divided by 10^{45} (1 followed by forty-five zeroes).... Finally, scientists estimate that the universe is about fourteen billion years old, meaning the universe is itself millions of times younger than 10^{25} seconds. [Fourteen billion years is roughly 10^{18} seconds, so by choosing 10^{25}, Dembski pushes his bound about 10 million times higher, again in the interest of bending over backwards to avoid false positives.] If we now assume that any physical event in the universe requires the transition of at least one elementary particle (most events require far more, of course), then these limits on the universe suggest that the total number of events throughout cosmic history could not have exceeded 10^{80} x 10^{45} x 10^{25} = 10^{150}. This means that any specified event whose probability is less than 1 chance in 10^{150} will remain improbable even if we let every corner and every moment of the universe roll the proverbial dice. The universe isn't big enough, fast enough, or old enough to roll the dice enough times to have a realistic chance of randomly generating specified events that are this improbable.[11]

The universal probability bound is clearly a very small probability. But if we start applying it to key molecular biological structures such as proteins (discussed in Chapter 2), we see that the chances of their occurrence are much smaller. Hence Dembski's filter would bring back a verdict of intelligent design. The filter would also support Murray Eden's surmise that proteins with 200 amino acids or more are not going to be products of chance. Eugene Koonin calculated the odds of abiogenesis as 1 in $10^{1,018}$, and he implicitly endorses the logic of Dembski's universal probability bound. That's why, to save a materialist conception of life, he posits an infinity of unobserved and, by definition, unobservable other universes, and then argues that ours is simply one of the lucky ones that won the freakishly improbable lottery of life. Koonin, in other words, overcomes the unimaginably long odds against abiogenesis by expanding the probabilistic resources to infinity, leading him to argue, in essence, that nothing is too improbable to occur.

One feature of Dembski's filter is that it sometimes rejects real design. This can happen in either of two ways. One way, as noted, is when a designed thing isn't complex/improbable enough to clear the absurdly high improbability threshold Dembski has set. The other way it can overlook design is if there is a specification that goes undetected. For example, take a look at the pattern "nfuijolt ju jt mjlf b xfbtfm." At first glance, this looks as though it is gibberish produced by monkeys, with no apparent specification and thus apparently not designed. But once we shift each letter down the alphabet by one—n becomes m, and so forth—we get "methinks it is like a weasel." There are, of course, much more sophisticated encryption techniques. Those who are trying to hide something work very hard to make the code indistinguishable from gibberish produced by random processes. But that takes intelligence. Intelligence can simulate randomness. But randomness cannot produce specified information—at least not of any great complexity.

Finally, the reader may have wondered how one applies the design filter to patterns and objects whose form is mathematically intractable. For example, how does one calculate the probability that Mount Rushmore was formed through natural processes? Here one does not require the rigor of the probability bound because it is easy to see that the monument's four carved faces are unimaginably improbable, so close is the fit of the rock to that of human faces, even for someone who had never seen the four US presidents represented there. In cases such as this, our intuitive sense of what needs a designer and what does not suffices. We do not need to do any elaborate calculations. Mount Rushmore, a book, and a watch are clear examples of intelligent productions, well before any calculations are attempted.

The problem with jumping from a watch to biology is that watches do not self-reproduce with occasional random mutations, so one cannot appeal to the Darwinian mechanism in the case of watches. One can appeal to the Darwinian mechanism in the case of living organisms. Whether the appeal is cogent or not is what Dembski's design filter, with its universal probability bound, is well-suited to adjudicate. But there is yet another difference between watches and living things: We know that watches are made by human beings, but no one was around when the first bacteria appeared on the face of the earth. Nor were any humans around when sea animals first appeared as part of the Cambrian explosion. Moreover, even if there were human observers around at the time to see it, they might still argue about the cause, because God is invisible and He may have been using subtle secondary causes. So one might argue that our intuition of design fails when taken out of the realm of human artifacts.

But no, none of this invalidates the design filter as applied to living things, because the information content in biology can be quantified. Everyone agrees that there is information coded by the nucleotides in DNA, so given the length of a genome and the number of amino acid changes in going from protein A to protein B, one can calculate the odds that a blind evolutionary process could cause such a transformation. Because of this relative ease of calculation, molecular biology features heavily in discussions of ID.

This brings criticisms from those who would like a more Aristotelian approach to biology, such as Edward Feser, who see ID as threatening to turn the whole field of biology into mechanics.[12] In the Aristotelian view of reality, plants and animals and humans are specified through their substantial forms: the vegetative, animal, and human soul. The soul, in these terms, isn't simply a ghost trapped in a body. It is the form of the body—it is what makes a human body *human*. It is, metaphysically, prior to its parts. It seems clear that the DNA is not the soul. Although there is much information in the DNA, there is nothing there that makes a horse a horse or a man a man. It is not even a blueprint for life, as recent studies in morphology make abundantly clear.[13] (See Chapter 2.)

One can be sympathetic to the Aristotelian approach, without rejecting ID just because some of its adherents focus on molecular biology and use mechanistic metaphors. The whole may be greater than the sum of its parts, but the parts are nevertheless important. One mutated gene in an embryo can be lethal. Also, just because some ID theorists have focused narrowly on molecular biology, this has not prevented other ID theorists from taking a more holistic approach to the study of life. Discovery Institute Press recently published *Your Designed Body*, which delves into the many systems of systems of systems that make up the human body. And they have published a series of books by biologist Michael Denton, who is just as eager to find some laws of form as any committed Aristotelian might be. Certainly the Aristotelian analysis has nothing to fear from ID. And committed Thomists (who follow not just Aristotle but the Christian synthesis of Saint Thomas Aquinas) should see ID scholars as allies, not enemies.

Fields other than biology also lend themselves to the ID perspective. Michael Denton, for example, is fascinated by the chemical potential of the elements. The same elements which are essential to intricate processes in the cell also lend themselves to macroscopic structures at the human level and to still larger processes such as the nuclear heating of the earth's core, continental drift, and mountain-building. This indicates a great depth of design in the elements and in the universe as a whole, which cannot easily (or at all) be mathematically quantified.

Many people, even those who instinctively recoil from ID in biology, are impressed by the fine-tuning of the universe. The various constants from physics, such as the ratios between the forces of gravity, electromagnetism, and the strong and weak nuclear interactions, are just right for life, as are a host of other conditions stemming from physics.[14] Often these examples of fine-tuning are called anthropic coincidences, because it seems that the whole universe is designed for the arrival of man (*anthropos*).

One can get a sense of how improbable life would be by doing a calculus of variation on a given condition. Sir Roger Penrose, for example, says that the odds against the low-entropy state that must have existed at the beginning of the Big Bang to account for the universe's present degree of order is something like one in 10 raised to the power of 10^{123}.[15] Mind you, this isn't the unimaginably small odds of 1 chance in 10^{123}. Rather, it is 1 chance in the number 1 followed by 10^{123} zeroes. And it's important to note that Penrose is asserting this despite his stated unwillingness to embrace the design explanation for fine tuning.[16] These are simply what his calculations tell him.

Jay Richards described another example, the cosmological constant:

> Cosmological constant (which controls the expansion speed of the universe) refers to the balance of the attractive force of gravity with a hypothesized repulsive force of space observable only at very large size scales. It must be very close to zero, that is, these two forces must be nearly perfectly balanced. To get the right balance, the cosmological constant must be fine-tuned to something like *1 part in 10^{120}*. If it were just slightly more positive, the universe would fly apart; slightly negative, and the universe would collapse.[17]

There is no need to list all the many other examples here. There are many of them. One of the great unknowns is whether the various constants are independently adjustable. Right now, in the standard model of particle physics, there are seventeen constants that need to be empirically measured rather than derived from the theory. Many physicists would prefer a theory that specified these constants internally. But we do not have such a theory. So we must be content to do some calculations to ascertain what would happen, to give just two examples, if the force of gravity were a thousand times greater or if the relative electromagnetic force were ten times weaker. But we are not sure that we can just multiply the probabilities as if they were independent.

Nevertheless, the fine-tuning in physics impresses the educated more than the fine-tuning of biology, probably because Darwinism has stripped away the sense of wonder of life and there is nothing analogous to natural selection in physics to account for the fine-tuning. One might also speculate that the fine-tuning in physics is further removed from our experience. It may be accounted for by a deistic conception of God, who fine-tuned the laws and constants of physics at the moment of the Big Bang and then stepped aside, whereas life suggests a much more intimate and perhaps repeated involvement on the part of God.

Whatever the psychological motivation for responding so differently to the two cases of fine-tuning, it is inconsistent to recognize fine-tuning and design in the realm of physics and astronomy and to deny it in biology. Kenneth Miller is clearly impressed by the anthropic coincidences in physics and astronomy. He takes a few pages of his *Finding Darwin's God* to explain the power of this argument to design. In particular, he chides the atheist Daniel Dennett for looking to a myriad of multiplying universes to escape the anthropic implications of the one universe that we find ourselves in and can actually observe. "Dennett knows that we will never be able to find, even in principle, evidence for any of those parallel universes," writes Miller. "If they existed, we could neither communicate with them nor observe them. Nonetheless, he is willing to postulate their existence because it relieves us of the need to find another reason for the elegant 'anthropic coincidences' of our universe. To those who doubt his solution, he writes that a multiplying swarm of universes is at least as good an explanation 'as any traditional alternative.'"[18]

Miller rightly notes that Dennett goes to such lengths because he is wedded to his atheism and needs to escape from God. But where are the Darwinists' numerous intermediate forms in the fossil record? They do not exist, but somehow they must exist in the minds of those who are wedded to methodological materialism. Would that Miller took the time to examine his own prejudices.

In the end, the theory of intelligent design is about inference to the best explanation. Everyone agrees that life looks designed. There is no plausible explanation of how a mindless process could have designed life. Therefore the best explanation is that life was indeed designed by a designing mind. It is common sense, which unfortunately is in short supply in our present culture.

Intelligent Design—Criticisms from Mainstream Science

Most Darwinists have not been convinced by Dembski or Behe. This is not surprising. The history of science clearly shows that scientists long committed to a paradigm are loath to give it up, even when the evidence has turned against it. Committed Darwinists, by and large, would like to rid the world of any challenge to their materialistic system, and to that end, several of their spokespersons paint ID arguments as biblical creationism masquerading as science.

A recent example of that is the apology issued by the *Journal of Theoretical Biology*, which published a peer-reviewed article on fine-tuning in biology in September 2020. A month later they published the following apology: "Since the publication of the paper it has now become evident that the authors are connected to a creationist group (although their addresses are given on the paper as departments in bona fide universities). We were unaware of this fact while the paper was being reviewed."[19] The paper had no theology in it. And it did not attempt to hide an association with ID thinkers, obvious from the fact that the authors cited well-known proponents of ID. It had been accepted on its scientific merit. But then the Darwinian thought police got into the act and the journal immediately caved.

Perhaps, in its defense, the *Journal of Theoretical Biology* could point to the Wikipedia article on ID: "Intelligent design (ID) is a religious argument for the existence of God, presented by its proponents as 'an evidence-based scientific theory about life's origins,' though it has been discredited as pseudoscience." For this entry, Wikipedia was awarded the Censor of the Year prize by Discovery Institute's Center for Science and Culture.[20] But of course one would expect the hub of the intelligent design community to complain about such a characterization. However, they are not the only ones complaining about such bias. One of the cofounders of Wikipedia, philosopher Larry Sanger, has issued a similar complaint:

As the originator of and the first person to elaborate Wikipedia's neutrality policy, and as an agnostic who believes intelligent design to be completely wrong, I just have to say that this article is appallingly biased. It simply cannot be defended as neutral.... I'm not here to argue the point, as I completely despair of persuading Wikipedians of the error of their ways. I'm just officially registering my protest.[21]

Darwinian zealots go far beyond grossly mischaracterizing intelligent design. They also have gone after the careers of ID theorists. Discovery Institute provides many examples of the injustices that ID proponents have faced: Scott Minnich, Richard Sternberg, Gunter Bechly, Dean Kenyon, Guillermo Gonzalez, and others.[22] Given what the Darwinists in academia are willing to do to their colleagues, it is no surprise that the list of around one thousand scientists who have so far signed the declaration at dissentfromdarwin.org includes few who have yet to get tenure. As one visiting scientist from China, J. Y. Chen, an expert on Cambrian fossils, put it, "In China we can criticize Darwin, but not the government; in America, you can criticize the government, but not Darwin."[23]

It may be helpful to give a few more details about a particular case of harassment by the Darwinist enforcers. It happened to Eric Hedin in 2013 and is described in his book *Canceled Science*. He was a tenure-track professor of physics at Ball State University in Indiana, teaching an honors course called The Boundaries of Science, in which he explained the state of the science in various fields and discussed with his students the philosophical implications of the technical research. He had been teaching the course for six years, and it was an immensely popular class, as attested by letters of support from students of various religious beliefs, as well as from students with no religious beliefs. These former students wrote in his defense after the attack started. He describes the first intimation of the controversy:

> Jerry Coyne, a nationally known atheist, blogger, and evolutionary biologist, had acquired a copy of my class syllabus and asked my chair to verify if it was being used in a science class at Ball State. Dr. Coyne opined that the course amounted to teaching a

"religiously infused science course" at a public university and violated the separation of church and state. On his view, the course was simply "religion served under the guise of science."[24]

Ball State complied with Coyne's request and sent him the syllabus. The Freedom from Religion Foundation also advocated against the course and Hedin. The controversy made it into the news. There were all sorts of misrepresentations, which *Canceled Science* details, with thorough references. In the end, the course was indeed canceled, although Hedin was retained as a professor at Ball State. The whole thing was so distasteful that even Larry Moran, a biochemist from the University of Toronto who often writes against ID, wrote on his blog *Sandwalk*: "I ban people from Sandwalk if I ever hear of them trying to intimidate someone by complaining to their employer. That's unacceptable behavior in my book."[25] Unacceptable to decent people interested in the truth and common decency, yes, but it happens all the same.

It is true that some proponents of ID are also young-earth creationists (YEC). And it is true that many have religious motivations for their advocacy. And yes, for many, myself included, ID provides a natural support for the act of faith in God. But using these facts to attack and dismiss ID involves a well-known rhetorical fallacy—attacking the person rather than rebutting the argument. Anytime you see such a move in a debate, ask yourself: If the argument under review is as weak as the critic says, why doesn't he rebut the argument itself instead of switching to a personal attack involving accusations that are beside the point?

Imagine if it had come to light that the scientist who first made the case for continental drift was a raging puzzle fanatic. Loved puzzles. Couldn't get enough of them. Now imagine that he and his chief opponent agreed to a public debate, and in that debate his opponent dismisses his evidence-based case by saying, "Of course you think the continents drift. Of course you think South America and Africa once fit neatly together like two puzzle pieces. You are a notorious puzzle fanatic, so naturally you want to treat the continents as puzzle pieces that fit together." Even if this early proponent of continental drift truly were a raging puzzle fanatic, it would be beside the point. Moreover, the fact that his opponent chose to attack the man instead of mounting an evidence-based argument for continental stasis should trigger the baloney detector of every thoughtful person in the audience.

Biases cut both ways. Perhaps the inclination to reject the evidence for design in nature is explained by an *odium Dei*: hatred of God. An interesting notion, but there remains the actual evidence pro and con for design. Any sort of a constructive debate will have to look at the evidence, rather than resorting to questioning the motives of one's adversary.

The ID argument does not claim to *prove* the existence of God as He is revealed in the Bible. It merely concludes that there is design in nature that cannot be accounted for by the (putatively) blind processes of physics or chemistry. Perhaps it is not surprising that two atheists who have written against Darwinism are both philosophers of the mind. Jerry Fodor has already been cited in the introduction. Here it will be useful to quote at length the discussion of Thomas Nagel, from his book *Mind and Cosmos*, which has for its subtitle *Why the Materialist Neo-Darwinian Conception of Nature Is Almost Certainly False*. He writes:

> Even though writers like Michael Behe and Stephen Meyer are motivated at least in part by their religious beliefs, the empirical arguments they offer against the likelihood that the origin of life and its evolutionary history can be fully explained by physics and chemistry are of great interest in themselves. Another skeptic, David Berlinski, has brought out these problems vividly without reference to the design inference. Even if one is not drawn to the alternative of an explanation by the action of a designer, the problems that these iconoclasts pose for the orthodox scientific consensus should be taken seriously. They do not deserve the scorn with which they are commonly met. It is manifestly unfair.
>
> Those who have seriously criticized these arguments have certainly shown that there are ways to resist the design conclusion; but the general force of the negative part of the intelligent design

position—skepticism about the likelihood of the orthodox reductive view, given the available evidence—does not appear to me to have been destroyed in these exchanges.... To anyone interested in the basis of this judgment, I can only recommend a careful reading of some of the leading advocates on both sides of the issue.[26]

An excellent place to start such reading is Michael Behe's *A Mousetrap for Darwin* (2020). In this volume, Behe collected over one hundred previously published essays in which he directly engaged with his critics. It is instructive reading because it is clear that the critics fail to engage Behe's arguments, or they deliberately misstate them, or they resort to childish name-calling. In replying to them, Behe shows the patience of Job. For example, in "Drawing to a Close with Moran" (Essay 78), Behe quotes Moran as saying, "His sycophants are promoting the idea." Behe responds, "Name-calling is puerile." Elsewhere Moran commented, "This is hard for IDiots to understand." Behe: "See my remarks above about name-calling." And remember that Moran is one of the more decent ones.[27]

The idea that so offended Moran came from Behe's second major book, *The Edge of Evolution*. In the first book, *Darwin's Black Box*, Behe presented various features of molecular biology of such integrated complexity that they could not be accounted for by means of a slow Darwinian process, but nowhere in the book did Behe say that natural selection can account for nothing. In *The Edge of Evolution*, Behe tried to get a sense of how much a Darwinian process alone could achieve.

Among the biological systems he analyzed was the ability of the *Plasmodium* parasite, which causes malaria, to develop resistance to the various drugs being used to fight it. It is an excellent test case, because there is a tremendous amount of data on malaria. Because the parasite is so numerous within the individual infected human body and the number of humans who have been infected so high, the number of these single-celled parasites that have had a chance to mutate is great enough that researchers can readily compare predicted probabilities of mutations with observed data.

It turns out that most anti-malarial drugs are not very effective, because a single mutation is all that is needed to give the parasite resistance to the drug. However, there was a very effective drug called chloroquine. The field data—not probability calculations—showed that the parasite hit upon resistance only about once every 10^{20} divisions. This is derived from the estimated number of parasites per patient, multiplied by the estimated number of patients suffering from malaria and treated with chloroquine, divided by the number of times that the drug ceased to work *de novo*. It turns out that this number, derived from field data, matched quite closely what Behe estimated would be necessary if two mutations rather than one were necessary to achieve resistance.

If, for example, the mutation rate is 10^{-10} per nucleotide and if two nucleotides are needed to change, then the probability becomes 10^{-20} of getting the right combination, which matches the field data quite nicely. Admittedly this is simplified. If, for example, changing one nucleotide gives some advantage to the parasite, natural selection can reduce the waiting time for the second mutation. There are other objections too, which, at the time of writing, Behe anticipated but could not answer because some biological details of the parasite had yet to be analyzed. But he presented what data he had as indications that the waiting times for natural selection to hit upon a significant advantage that could eventually lead to diversification of life were far too long to make random genetic mutation and natural selection a viable basis for a grand theory of evolution. At the time, he speculated that the limit of such evolutionary change, the "edge of evolution," was somewhere between species and class.[28] Fish, amphibians, reptiles, birds, and mammals are classes within the phylum Chordata. So Behe's estimate was somewhat generous towards the Darwinian mechanism, but refuted the grand claim that the mechanism had evolved all of life's diversity beginning from a universal common ancestor.

When *The Edge of Evolution* was published, the scientific establishment's two most prestigious journals, *Nature* and *Science*, each gave it long and negative reviews to reassure the Darwinist faithful that natural selection was omnipotent. The interested reader can easily find these reviews and Behe's response to them. Some of the discussion is technical, but an intelligent reader can easily pick up that Miller in *Nature* and Carroll in *Science* were not really engaging with Behe's arguments.

I will discuss the case of Kenneth Miller in Chapter 8, but here one can get a sense of the kind of argumentation he used, from Behe's response:

> When Miller writes of "protein-to-protein" binding sites in one sentence, wouldn't you expect the papers he cites in the next sentence would be about protein-to-protein binding sites? Well, although the casual reader wouldn't be able to tell, they aren't. *None* of the papers Miller cites involve protein-protein binding sites.... What's more, none of the papers deals with evolution in nature. They all concern laboratory studies where very intelligent investigators purposely rearrange, manipulate, and engineer genes (not whole cells or organisms) to achieve their own goals....
>
> Miller's snide comment, that apparently I hadn't followed these developments, seems pretty silly, since it's so easy to find that I followed them closely. You'd think he should have noticed that I cited the *Annual Reviews* article in *The Edge of Evolution* in Appendix D, which deals in detail with Wendell Lim's interesting work on domain swapping. You'd think that he easily might have checked and seen that I was quoted in the *New York Times* commenting on Joseph Thornton's *Science* paper when it first came out a year ago. You'd also think that he'd then have to tell readers of the review why I thought the papers weren't pertinent. You'd be thinking wrong....
>
> Call it the principle of malignant reading. He's been doing it for years with the arguments of *Darwin's Black Box*, and he continues it in this review. For example, despite being repeatedly told by me and others that by an "irreducibly complex" system I mean one in which removal of a part destroys the function of the system itself, Miller says, no, to him the phrase will mean that none of the remaining parts can be used for anything else—a straw man which can easily be knocked down. Unconscionably, he passes off his own tendentious view to the public as mine. People who look to Miller for a fair engagement of the arguments of intelligent design are very poorly served.[29]

Reviews such as Miller's give substance to Nagel's observation that ID proponents, while deserving to be taken seriously and to have their arguments debated on their merits, are instead met only with scorn by the leading defenders of Darwinism.

Behe's next book, *Darwin Devolves*, also received its share of scorn from prestigious quarters. But that is hardly surprising, since he revised his estimate for the edge of evolution, suggesting that he had overestimated its reach in his previous book and that additional research data since then suggested that the Darwinian evolutionary mechanism could reach, more modestly, to somewhere between family and genus, rather than extending as far as class. Below the class of mammals, for example, is the order Carnivora, which contains the family Canidae and the family Felidae. So at the time of *The Edge of Evolution*, Behe thought that cats and dogs and even cows might have been derived from some common ancestor via natural selection. In *Darwin Devolves*, it became an open question for Behe whether natural selection could evolve even cats and dogs from a common ancestor. This is based on more recent biological data suggesting that evolution is limited because diversification of species (of finches, bears, dogs, cichlid fishes, and other plants and animals besides) is usually accompanied by some genetic degradation. Devolution, Behe argues, is a poor path to generating entirely new biological families, much less new orders. If you pile up too much devolution, it eventually takes its toll, and in any case, the breaking of genetic parts is no way to build up entirely novel forms.

Intelligent Design—Young Earth Criticisms

Michael Behe might be the best-known proponent of ID, but not everyone who thinks that Darwinian evolution has a limited capability is happy with his approach. Young-earth creationists (YECs), for example, do not countenance a universe that is billions of years old. And they rule out universal common descent, even one intelligently guided by God. They are convinced that the clear teachings of the Bible require a young earth and special creation.

Many of these YEC proponents, it should be stressed, are far from the combative, Bible-thumping illiterates that they are commonly portrayed as in the mainstream media. Many of them are learned and cordial, happy to embrace the big-tent approach to the intelligent design community that Phillip Johnson urged.

Moreover, it's not as if they have nothing to complain about beyond the damning weaknesses in Darwinian theory. I am aware that physics is very far from giving a comprehensive view of all the inorganic aspects of the universe. Quantum mechanics and general relativity are at serious odds with one another. Attempts to unite them in some grand unified theory (GUT) or a theory of everything (TOE), usually string theory, have so far produced no testable results, as may be gathered from some recent titles by physicists in the know: *Not Even Wrong: The Failure of String Theory and the Search for Unity in Physical Law* by Peter Woit, and *Farewell to Reality: How Modern Physics Has Betrayed the Search for Scientific Truth* by Peter Baggott.

The average non-physicist will have a hard time appreciating the contradictions between quantum mechanics and general relativity, but anyone should see that there is a problem in cosmology if about 83 percent of the mass necessary to hold our galaxy together cannot be accounted for. Physicists now speak of dark matter and dark energy, but those are just shorthand for ignorance. A more candid physicist likened our present dark matter to Ptolemy's deferents and epicycles, which long ago were swept into the dustbin of intellectual history.[30] In light of this, is not physics, or more generally science, always going to be in flux, with today's certainties dismissed in the future as patent absurdities? Might not the Bible be a better source of truth?

The Catholic Church, of course, teaches that the Bible is the inspired word of God. But acknowledgment of that truth still leaves a great number of considerations on the table. Yes, the Bible and Tradition are the only sources for some truths: God is a Trinity of Persons; Jesus Christ is true God and true man; the Blessed Virgin Mary was conceived without sin; Christ is substantially present in the Eucharist. But there are many truths that the Bible and the Tradition do not speak to unequivocally, or do not speak to at all, truths that human beings can discover for themselves by carefully consulting God's other book, the book of nature: such truths as the Pythagorean theorem; that water is composed of hydrogen and oxygen; that most plants need sunlight to grow. There is, in sum, an autonomous sphere for human reason. And most of the questions of physics and biology legitimately fall into this sphere.

The problems arise when scientists make claims that contradict, or seem to contradict, the teachings of Scripture or Tradition. This can happen in either of two ways. First, someone can misinterpret Scripture so as to insist it is saying something it isn't. Someone might, for example, insist that the Bible teaches that the sun revolves around the earth because it speaks of the sun rising and setting. But this is a misreading of such passages, which are only describing what the sun does relative to human observers, using the language of everyday life. Second, scientists can draw a false conclusion about nature. For instance, scientists regularly insisted that the universe is eternal (despite a lack of evidence), while the Bible speaks of a beginning, and later scientific findings have strongly suggested that the cosmos did indeed have a beginning in time.

The Church teaches that God created *ex nihilo* and that time began with God's creating something changeable, since time is a measure of change and so does not apply to God. Moreover, the Church teaches that human beings are created in the image of God, with an immortal soul. It also says that we are all descendants of Adam and Eve. These teachings are found in the opening chapters of the book of Genesis, although other biblical texts can be adduced in their support.

These teachings are unambiguous and theologically central. At the same time, we can and should acknowledge that anyone who tries to read the first eleven chapters of the book of Genesis will encounter many difficulties. He will have to ask himself what sort of literature he is reading. Is it, for example, a morality tale, such as the story of the boy who cried wolf? Such an account does not set out to answer questions as to the boy's name or the exact location of his village. I sometimes dress up the story in my visits to primary school classrooms with such details just because I know that younger children will interrupt me to ask these questions. But no adult hearing my account would think that the village actually existed or would be bothered by my using a completely different name in recounting the story at a future date. The whole point of the story is to illustrate the fact that no one believes liars.

Are the earliest chapters of Genesis such a genre, or some other? Is the story of Adam and Eve's fall a parable, akin to the parables of Jesus, full of truth but not meant to be read as point-for-point history? Or perhaps those early chapters are *poetic history*, an account containing historical facts but presented in a stylized and sometimes highly compressed form, the better to highlight key theological truths, fit it into the limited space of a scroll, and render it easier to memorize for an ancient society in which only a small fraction of the population could read and access rare and expensive scrolls. Or perhaps a more accurate term would be *exalted prose narrative*, the preferred term of the evangelical Old Testament scholar C. John Collins,[31] who eschews a young-earth reading of the early chapters of Genesis while still maintaining that Adam and Eve were actual historical figures, created directly by God.

Christian thinkers with a high view of scripture as diverse as Augustine, Thomas Aquinas, Bonaventure, and C. S. Lewis have pondered how to read the opening chapters of Genesis, and not because they were attempting something heterodox or illicit. The material itself invites such questions. Genesis 1 speaks of days one, two, and three, of evening and morning, all before the sun and moon are created on day four. Yes, God could have employed a temporary divine light to create the rhythm of night and day for a spinning Earth before the creation of the sun—an awkward and inelegant approach but not beyond God's power. However, isn't it at least as reasonable, given the many genre forms in the Bible, to suppose that having the sun not appear until day four might be the passage's way of signaling to the reader that this section of Genesis is not referring to ordinary 24-hour earth days?

To ask what genre a particular section of Scripture is does not deny that it is the inspired word of God. It only allows for the possibility that we may need to fine-tune our understanding of it. To ask about genre is exactly what one should do if one is convinced that the material in question is the inspired word of God. For to ask the question is to seek to read the text as carefully as possible, and what text merits greater care?

Clearly the book of Genesis is important to the Christian faith. It is about God taking the initiative in creating the world. God is the uncreated source of all being. The book speaks of a beginning rather than of an eternal universe. And there is a progression from plants to animals to humans, who alone are created in the image of God, and are given the mission to multiply and have dominion over everything God had created. Further, the first chapter reveals that what God has created is good, with no admixture of evil. Finally and, if Fr. Stanley Jaki is to be believed, perhaps most importantly, the chapter stresses the importance of the Sabbath rest as an essential characteristic of the Jewish people, which set them apart from their neighbors who also measured time in periods of seven days.[32]

There is certainly more than the Sabbath day message to the first chapter of Genesis. But mentioning a progression from plants to man might alone be enough to be suspected of concordism. Jaki, for example, examined many commentaries on Genesis across the centuries, which tried to square the biblical text with contemporary science, and he found them all wanting. He suggests that the naming of the various created beings is just a way of emphasizing the whole by enumerating its parts.

There are many ways to read Genesis, and Christians have tried different approaches over the centuries. The default tendency seems to have been to see the days of Genesis 1 as twenty-four-hour periods, but no orthodox Christian thought that because God is said to have walked through the Garden of Eden that he had a physical body or that the devil was a talking snake. Saint Augustine did not think that the days of creation were our twenty-four-hour time divisions, and he was not censured for this view.

It would take us too far afield to trace the origin of the more extreme version of *sola scriptura*, an outlook that makes it all the easier to take "day" to mean twenty-four hours and to regard the Flood as universal. If such an understanding is now found mainly in Protestant circles, most Catholics, until quite recently, shared these beliefs. Saint Nicholas Steno (1638–1686), widely considered one of the founders of geology, believed that Noah's Flood was responsible for the rock formations around Florence.[33] The same approach to the Bible which encouraged and guarded the twenty-four-hour understanding of "days" and the conviction that Noah's Flood covered the whole earth naturally lent itself to thinking about other questions, such as the age of the earth. The Bible provides genealogies, which Bishop Ussher analyzed in the seventeenth century to pinpoint the exact time of the first creation as around six o'clock in the evening of October 22, 4004 BC. Ussher was an Anglican, but most Christians adopted his dating. My 1899 edition of the Douay-Rheims (Catholic) Bible has his chronology in the margins. But by then, scientific findings, especially in geology, were casting serious doubt on such a young earth.

There are many good reasons to think that the universe is older than some six thousand years. There is the Big Bang theory, which was first posited by Alexander Friedmann in 1922 and shortly thereafter independently developed by the Catholic priest and scientist Georges Lemaître, whose work brought it into the mainstream of physics. The discovery of the 3K background radiation in 1964 tipped the scales heavily in favor of the Big Bang theory over the competing steady state theory, which held that the universe was eternal. So now most scientists accept that the universe had a beginning, and an age of some 13.8 billion years, from the physics of the Big Bang. As for Earth, the astronomy of the solar system, coupled with geological considerations based on radiometric dating, has homed in on an age of approximately 4.5 billion years.

It should again be emphasized that those who want to hold on to a young universe and young earth are not ignorant people. Among them are scientists with doctorates in diverse fields. Ultimately, one can grant that God could have created the universe some six thousand years ago, with fossilized dinosaur bones covered by many layers of rock via the Great Flood, alternating magnetization in matching rock formations on the two sides of the mid-Atlantic rift laid down with great rapidity (quite differently from how they are laid down today), and light from distant galaxies somehow reaching us in thousands of years rather than in the millions and billions of years indicated by their great distance and the speed of light. But why would God lay down so many misleading clues? Is it just to teach us the humility to read the book of Genesis in a literalistic way, never mind these seemingly contrary clues?

Yes, one can come up with ways to explain away how some of the mainstream dating systems do an impressive job of confirming each other, and one can point to instances where dating attempts have gone badly awry. From there one can argue that the various dating systems are wildly misleading and the young-earth reading of Genesis correct. Those making such arguments know their physics and can point out scientific assumptions that are made without proof. But their arguments strike me as *ad hoc* rebuttals rather than successful efforts to form a more comprehensive scientific picture.

I am not going to delve into the details of any particular dating controversy here, for my strong sense is that the young-earth program is fueled by a search for answers in Genesis that the book was never meant to provide. I know there are thoughtful Christians who are persuaded that letting go of the argument for a young earth is a mistake not only substantively but also rhetorically, sacrificing as it does a line of argument that, if confirmed, would discredit modern evolutionary theory beyond a shadow of a doubt. (After all, nobody believes Darwinian evolution could do all the necessary creative work required of it in a mere 6,000–10,000 years.) While I can appreciate the attraction of such a knock-down argument, were it true, I must demur simply because I am not convinced that the argument is true. I respectfully disagree that the Bible's opening chapters were meant to convey the claims about creation at the core of the young-earth interpretation, and I am convinced that God's book of general revelation, nature, points strongly to God's having created the universe *ex nihilo* some 13.8 billion years ago.

I also see a rhetorical dimension to this matter: Science was born under and nurtured by the Christian faith, primarily if not exclusively by the Christianity of Medieval and Renaissance Europe. Forcing science to support a very specific understanding of Genesis, in the face of so much contrary scientific evidence, will just encourage many scientists to continue in their rejection of a key foundation of the scientific revolution, the Judeo-Christian understanding of God, man, and creation, and thus reject the profound wisdom of the Bible.

To be clear, it does not follow from embracing an old universe and the Big Bang model that everything needs to take place according to the laws of physics. God can act directly in nature, and is not bound by natural laws. But it seems that even when He performs a miracle, He sometimes opts to work with the laws and forces of nature. In the book of Exodus, for example, we find this: "Then Moses stretched out his hand over the sea; and the Lord drove back the sea with a strong east wind all night long and turned the sea into dry ground. The waters were split, so that the Israelites entered into the midst of the sea on dry land, with the water as a wall to their right and to their left" (Exodus 14:21–22).[34] The miracle seems to be a combination of wind and something beyond nature. But why use wind at all? He did not have to. Perhaps it was simply because He preferred to.

Coming back to the question of evolution, there are various preferences among those who look beyond mindless processes to explain how the intelligent designer caused the various species of plants and animals to arise. Some, such as Michael Behe, are not bothered by universal common ancestry. God's Providence, he says, could have guided quantum processes to construct the irreducibly complex systems found in nature. We leave that to Him. All that science can do is to deliver its verdict that natural selection and any other mindless processes are not up to the job.[35]

Others see in the "kinds" of plants and animals mentioned in Genesis the limits of evolution. God made each of the kinds directly. Kinds are sometimes called "natural species" and are akin to the Linnaean classification of genus or family. Just as Behe restricted the power of natural selection to lie between genus and family, some Christians accept diversification within "kinds," but no further. Hence, for them, universal common ancestry is ruled out by the Bible.

As far as I can see, it is a matter of philosophical preference whether one accepts or denies universal common ancestry. The main reason for ruling it out is based on the best empirical evidence we have: bushes of life rather than a tree of life. On the other hand, if one likes the idea of the unity of life, one might say that God was involved at millions and even billions of steps along the way to help the creative evolutionary process along, but always doing so parent to offspring—a view we could call "intelligently designed universal common ancestry."

When it comes to man, one could take the line that since some philosophical traditions tend to view man as a microcosm, it is only fitting that we developed out of other animal life. On the other hand, if one takes seriously the arguments for human exceptionalism, to be discussed in Chapter 7, it is difficult to imagine how Adam and Eve would have arrived on the scene as the offspring of some speechless primates, whose task it would have been to nourish and educate them until the time God gave them the gift of speech. Moreover, there is the story of Eve being formed from the side of Adam, which has served as the basis of much theological reflection, and which it would be impossible to fit into an evolutionary scenario. I tend to agree with Sermonti's view: "My thesis all along has been that Man was born all of a sudden, in a great leap—i.e., in non-Darwinian fashion. His ontological leap was a biological leap as well."[36]

The Permeability of the Physical Universe

The Marxist biologist Richard Lewontin may want to keep the Divine Foot from entering into science, but no serious Christian can countenance that. Ultimately, every Catholic must grant that science cannot supply a comprehensive view of nature. This follows from the revealed truth that the human soul is not an epiphenomenon of matter. Pius XII acknowledged the possibility that the human body may have arisen from preexisting life, but he reminded the faithful that the "Catholic faith obliges us to hold that souls are immediately created by God."[37]

This is because the human soul is a spiritual reality. No amount of shuffling and modifying matter can give rise to a spiritual being.

This is because "spiritual" means "immaterial"—that is, not made of matter. (More on why the soul must be immaterial in Chapter 7). Material beings do not need to be created *ex nihilo* by God. He can use existing matter, just as I can disassemble a brick wall and use the bricks to make a fancy barbeque. But granted that the soul is immaterial, it is not separable into parts and cannot be fashioned from parts. It needs to be created *ex nihilo*. And God alone possesses the power to create *ex nihilo*.

Why? A full explanation would require something approaching a short course in Thomistic metaphysics, which would take us well beyond the scope of this book, but perhaps the following will give the reader some idea of the argument.[38]

An instrumental cause needs to contribute something of its own nature for it to participate in the chain of causality. If I cut down a tree, I am the agent. I am primarily responsible. But since I cannot cut down a mature oak tree with my bare hands, I need to use an instrument—a saw or an axe. The saw or the axe can both be instrumental causes here because they have sharp edges. Sharpness is an aspect of their natures; therefore, I can use them for cutting.

We can look to another example. A mute president can make a speech to Congress by asking someone with intact vocal cords to read out his text. Whoever reads his text is the instrumental cause of the message being delivered. There are many who can be such instruments, but clearly an elephant could not. It lacks the relevant something to contribute of its own nature to the delivering of the speech.

God creates the natural world and holds it in being. He gives it laws to follow. So it is legitimate to say that God is active in everything that happens in the universe. More particularly, believers can also recognize God's more direct actions in salvation history, such as the parting of the Red Sea or His messages spoken by His prophets. In the first example, the wind is His instrument; in the second, His prophets. So now we return to the claim above, that God can create *ex nihilo*. How might we reasonably conclude this, and if He can create *ex nihilo*, do we have reason to believe He does so in the creation of human souls?

We saw above that if we grant that the soul is immaterial, then it needs to be created *ex nihilo*. It cannot be pieced together from existing parts, as one might do for, say, a robot. We also saw that in order to create *ex nihilo*, to bring into existence from nothing, the agent must have existence as its essence. After all, such an agent in such a case is imparting being to a creation, not just form. This is why Saint Thomas said that "God's essence is His existence." Less obscurely, we can say that God necessarily exists—he can't not exist. He exists in every possible world. He alone is the necessary Being, the uncaused Cause. And since He alone has the power to create *ex nihilo*, He alone can create immaterial souls, since they can only be created *ex nihilo*. This, of course, is a very difficult concept. All those who mockingly ask, "If everything has a cause, who caused God?," clearly do not understand. It is more precise to say, *everything that has come into being, has a cause*. But of all that is, was, or ever will be, God alone did not come into being. He exists eternally. Existence is part of His essence—intrinsic to Him necessarily and, thus, eternally.

What then? If the soul indeed is a spiritual (an immaterial) reality, and millions of new souls are created each year, then the material world must be permeable to the action of God.

But that is not the only reason for the permeability of creation. Each human being who attains the age of reason can make free choices that often result in physical effects: moving an arm, speaking, chopping down a tree. Physics, therefore, must leave the world open to accommodate our freedom. Physics cannot fully describe our actions.

The basic intuition that we are free agents who can cause physical effects is not part of the Newtonian model of the universe. As a simplified working model of nature for doing physics, the Newtonian model need not lead to mischief. The problem comes when scientists began treating it as something more than a simplified model of nature.

"Laplace's demon" became the shorthand reference to the problem of physical determinism. This hypothetical being bears the name of the French mathematician and physicist Pierre-Simon Laplace (1749–1827), who introduced him to the world to make a philosophical point. Laplace's demon was a master calculator. The claim is that if he were to know the mass, position, and speed of every particle in the universe at a given instant, then this hypothetical demon (who was outside the system) could foretell the future state of the universe by solving the resulting equations of Newtonian physics; and he could know the past by solving them backwards. He could put both fortune-tellers and historians out of business.

No one, of course, believed that this was possible in practice. But many were puzzled as to how man could be free in a Newtonian universe; so they went on to propose many ingenious solutions. Some invoked God's omniscience of future human choices which He used to set the initial clockwork, i.e., the Newtonian universe, just right to anticipate all the future choices. Others focused on mathematical functions which gave multiple solutions. A lot of ink was spilled throughout the nineteenth century wrestling with this problem.

The problem partially resolved itself with the arrival of quantum mechanics, which made it impossible to know both the starting positions and momenta of each particle and could only make probabilistic predictions as to their future interactions. But the problem remained as to how a human being could make a free decision. The probabilistic nature of quantum mechanics made it possible for a man to move an arm or not without violating the laws of physics—something impossible under Newton. But actions determined by the collapse of a probabilistic wave equation are hardly the actions of a free human being. True human actions must originate in the soul and not in a wave function, whatever that may be.

What is essential is that the spiritual soul be able to alter the world described by physics and chemistry. So the cosmos needs to be permeable to the action of spirits. And it is not just God and the human souls that act. There are also angels and demons. The whole visible creation is meant to be a stage where persons can interact and come to know and love one another and God. In the words of the prophet Isaiah, God "established it; he did not create it a chaos, he formed it to be inhabited!" (Isaiah 45:18). Physics and chemistry capture only a part of God's creation.

This permeability opens up a whole new vista. It makes room for miracles. And it allows us to take seriously the various Gospel passages in which Christ expels demons from various individuals. The usual modern understanding of these events—when they are not dismissed as outright fabrications—is to insist that the afflicted were epileptics and the fit came to a natural end or that the sufferer was shocked by Christ's strong personality into coming out of whatever mental state he was in. It is a rather condescending *eisegesis*—a reading into the text—which belittles the ability of Christ's contemporaries to recognize true possession. The Pharisees were closer to the truth when they accused Christ of casting out devils by the power of Beelzebul.

Ridiculing the very idea that demons exist also makes nonsense of the various lives of the Saints, which specifically mention angels and demons. Some of these hagiographies can be dismissed as pious fabrication by overly credulous believers. But others are clearly not, such as Saint Catherine of Siena, the Curé of Ars, and Padre Pio.

In sum, Catholics cannot believe that the universe is impermeable to the action of spiritual beings and still remain Catholic.

God of the Gaps

One of the often-encountered objections to ID is that it is just a modern version of the "god of the gaps" fallacy. When faced with mysterious phenomena such as lightning, thunder, or earthquakes, the pre-Socratics explained them as the work of various deities: Zeus throwing thunderbolts or Poseidon ("The Earthshaker") hitting the ground with his trident. As understanding of nature progressed, these deities were swept into the dustbin of history. It can even be argued that science could progress only once such deities were banished from the mental landscape.

ID is thus portrayed as a science-stopper. Right now, the argument goes, we do not know how life began, but if we say that God did it, we will never know the "real" answer. So let's not give up on chance and necessity. At some point, we will hit upon some purely naturalistic explanation.

The god-of-the-gaps objection does have some merit to it, but it does not rule out ID. The progress of science has dethroned a multitude of false gods. For that it should be commended. But it cannot banish the God of Abraham, Isaac, and Jacob. He is not a stop-gap explanation for a particular phenomenon. He is the First Cause and the ground of all being, the Creator by whose intelligence the cosmos is sustained. But in most events there are also secondary causes at work, which science tries to find and better understand.

We have just seen that it would be pointless to look for a secondary cause of a human soul, because God alone can create a spiritual being. Given that vast numbers of scientists for many years have been trying to figure out how the specified complexity necessary for life could have arisen without the intervention of an intelligent agent and failed miserably, it may not be daring and premature to conclude that intelligent design is the better explanation.

The methodological materialist will object that, having let God into the lab, scientists will grow lazy, throwing up their hands at every mysterious physical phenomenon and saying, "God did it." But history tells precisely the opposite story. It was because Christians believed that God, a rational lawgiver, was the Creator of the universe that science got started in the first place. (More on this below.)

Nor has the design perspective outworn its usefulness. The ID perspective can still guide science in fruitful ways. Take, for example, the evolving view of DNA. At first, most of it was dismissed as "junk" because only about 2 percent of the human genome codes for genes and because biologists, tutored by the Darwinian paradigm, assumed that these non-coding strands of DNA were just useless leftovers from Darwinism's trial-and-error process. Darwinists trumpeted all the non-coding DNA as proof that evolution is messy and undirected, hardly the work of a wise designer. But as we saw in Chapter 2, design theorists pushed back, predicting that much of the non-coding DNA would turn out to have function. And that prediction turned out to be correct. In this case, then, the theory of intelligent design proved to be the more fruitful heuristic, with Darwinian presuppositions holding back scientific progress.

Fortunately, even in today's climate of prejudice against ID, its proponents sometimes manage to get their papers published in peer-reviewed journals. As of May 2024, there were over 200 such papers.[39]

This is good news, for science originated in Medieval Europe in part because Europe's cultural presuppositions at the time included the design perspective. The predominantly Catholic culture of Western Europe in the Middle Ages held that the universe was well-ordered because it was created by God, and that humanity could come to appreciate this order because we are made in God's image, which means we have the rationality to be able to investigate His creation.

Further, the Scriptures tell us that God "hast arranged all things by measure and number and weight" (Wisdom 11:20) and that "from the greatness and beauty of created things, comes a corresponding perception of their Creator" (Wisdom 13:5). Investigating the created world, then, was recommended as a good thing. Lest these verses be dismissed as coming from a book that is not even considered part of the Bible by Protestants, a survey of medieval literature found that one of the most commonly quoted biblical verses in medieval times was Wisdom 11:20.[40] It was clearly the inspiration for the famous work of medieval art shown in Figure 3.2. And as we will see in the next chapter, the insight conveyed in Wisdom 13:5 is also echoed in the New Testament by the apostle Paul.

The notion that Christendom gave rise to science is something foreign to most people. The average person has been taught that the Church's relationship to science boils down to the Galileo case, and not even the actual Galileo case, which is full of complex characters and nuances, but a cardboard caricature of that conflict, with nuances assiduously filtered out to prop up the "science vs. Christianity" myth. But historians of science know better, and in fact, it is conventional wisdom among historians of science that the thoroughly Catholic culture of medieval Western Europe was absolutely central to the rise of science.

Even non-Christians such as Alfred North Whitehead attest to the historical connection between the Church and the rise of science.[41] He comments:

> I do not think, however, that I have even yet brought out the greatest contribution of medievalism to the formation of the scientific movement. I mean the inexpugnable belief that every detailed occurrence can be correlated with its antecedents in a perfectly

Figure 3.2.
Frontispiece of Bible Moralisée. "God the Geometer." Artist unknown, circa 1220–1230.

definite manner, exemplifying general principles. Without this belief the incredible labours of scientists would be without hope. It is this instinctive conviction, vividly poised before the imagination, which is the motive power of research:—that there is a secret, a secret that can be unveiled. How has this conviction been so vividly implanted in the human mind?

When we compare this tone of thought in Europe with the attitude of other civilisations when left to themselves, there seems but one source for its origin. It must come from the medieval insistence on the rationality of God, conceived as with the personal energy of Jehovah and the rationality of a Greek philosopher. Every detail was supervised and ordered; the search into nature could only result in the vindication of the faith in rationality. Remember that I am not talking about the explicit belief of a few individuals. What I mean is the impress on the European mind arising from the unquestioned faith of centuries.[42]

Many of the ID proponents today who are producing good science are in line with this tradition and their success is a cause for hope.

Darwinism is not just a science-stopper; it is also, as Oxford mathematician and philosopher John Lennox points out, guilty of itself adopting the god-of-the-gaps fallacy. In this case, the god is natural selection. Whenever materialism cannot come up with any empirically verifiable explanation, it invokes natural selection. Lennox cites Nobel laureate physicist Robert Laughlin:

> Much of present day biological knowledge is ideological. A key symptom of ideological thinking is the explanation that has no implications and cannot be tested. I call such logical dead ends anti-theories because they have exactly the opposite effect of real theories; they stop thinking rather than stimulate it. Evolution by natural selection, for instance, which Darwin conceived as a great theory, has lately come to function as an anti-theory called upon to cover up embarrassing experimental shortcomings and legitimize findings that are at best questionable and at worst not even wrong. Your protein defies the laws of mass action—evolution did it! Your complicated mess of chemical reactions turns into a chicken—evolution! The brain works on logical principles no computer can emulate? Evolution is the cause![43]

This chapter and the one before it are, of course, a quick flyover of the case for intelligent design over against modern evolutionary theory. I would encourage the interested reader to delve deeper by reading Jerry Fodor and Massimo Piattelli-Palmarini's *What Darwin Got Wrong* for the negative case against Darwinism, and for the positive case for intelligent design in biology, Michael Behe's *Darwin Devolves*, Stephen Meyer's *Darwin's Doubt*, and Michael Denton's *Evolution: Still a Theory in Crisis*. Fodor and Piattelli-Palmarini are pretty sure that God does not exist but seem equally sure that Darwinism is dead—or that it ought to be dead, given the nonsense that it inspires in people in the affiliated fields of philosophy of mind and evolutionary psychology.

Behe and Meyer believe in the God of the Christians. Denton is also a supporter of ID, but he hopes to account for the structures of life through some as yet undiscovered laws of form, built into nature by the designer from the beginning, rather than by appealing to the actions of a designing intelligence at various points along the way in the history of life. He describes himself as "not in any sense a fervent believer in a God, or the Christian God." At the same time he doesn't describe himself as an atheist, but rather as being "on the edge of skepticism about theism."[44]

One might object that my reading list is rather skewed against Darwinism. I do not apologize for this, because our culture is saturated with pro-Darwinian propaganda. It is taught as settled science in schools. Every nature show throws in lines such as "the animal acquired this amazing trait over eons through evolution," without giving any idea of how evolution was supposed to have done it. You do not need to read a book by Richard Dawkins or Jerry Coyne to get a balanced view. You encounter their ideas every day.

An Intelligent Designer?

Some people's attitude toward intelligent design is shaped primarily by misinformation. They have been shown an off-putting caricature of it and backed away, with theists opting for an awkward marriage of theism and Darwinism as a seeming best resort. But I submit to you that in many cases, the negative attitude toward intelligent design stems from a negative attitude toward God. Although the logic of the design inference does not get us to the God of the Bible, the next obvious step is to ponder the nature of the designer. If it is a single being responsible, then we have in view a being capable of designing everything from laws and constants of physics and chemistry to the software in every living cell—a job description above the pay grade of even the cleverest of aliens.

The evidence from molecular biology was compelling enough to convert the renowned philosopher and notorious atheist Antony Flew to belief in God, albeit not to Christianity or to a belief in an afterlife. On the other hand, Thomas Nagel, although sympathetic to ID, ultimately rejects it. As he explains:

> I confess to an ungrounded assumption of my own, in not finding it possible to regard the design alternative as a real option. I lack the *sensus divinitatis* that enables—indeed compels—so many people to see in the world the expression of divine purpose as naturally as they see in a smiling face the expression of human feeling. (I am not just unreceptive but strongly averse to the idea, as I have said elsewhere.) So my speculations about an alternative to physics as a theory of everything do not invoke a transcendent being but tend toward complications to the immanent character of the natural order. That would also be a more unifying explanation than the design hypothesis. I disagree with the defenders of intelligent design in the assumption, one which they share with their opponents, that the only naturalistic alternative is a reductionist theory based on physical laws of the type with which we are familiar. Nevertheless, I believe the defenders of intelligent design deserve our gratitude for challenging a scientific worldview that owes some of the passion displayed by its adherents precisely to the fact that it is thought to liberate us from religion.[45]

Elsewhere, Nagel is, if possible, even more explicit about his bias against theism. "I want atheism to be true and am made uneasy by the fact that some of the most intelligent and well-informed people I know are religious believers," he commented. "It isn't just that I don't believe in God and, naturally, hope that I'm right in my belief. It's that I hope there is no God! I don't want there to be a God; I don't want the universe to be like that."[46]

One should be grateful for Thomas Nagel's candor and pray that God will give him the *sensus divinitatis*. Whether the *sensus divinitatis* is an aspect of human nature or whether it is something that only grace can bestow is better left to the theologians. But its lack in a person, combined with a positive desire for atheistic belief, is clearly problematic on one's path to God. Yet we can thank Nagel for pointing out that the passionate resistance to ID need not stem from true science but instead may arise from an inflexible commitment to atheism. In the next chapter, we will look at what implications accepting or rejecting ID has for establishing the existence of God.

4. Intelligent Design: A Preamble to a Powerful Way to God

*The sight of a feather in a peacock's tail,
whenever I gaze at it, makes me sick.*[1]
—Charles Darwin

*Our holy mother, the Church, holds and teaches that God, the first
principle and last end of all things, can be known with certainty
from the created world by the natural light of human reason.*[2]
—Constitution *Dei Filius*, Vatican I

Beauty is a powerful springboard to the divine. So it should come as no surprise that Darwin was sickened by the sight of a peacock's tail feathers. As we shall see in Chapter 8, he was a materialist from a young age, and determined to come up with a scientific justification for materialism. The glory of the peacock's tail was an affront to his theory of natural selection, which had no eye for beauty, only functionality. Darwin was certainly not interested in listening to an argument for the existence of God. And the feather must have engendered some inkling of the truth that he wanted to keep out of his heart.

The first Vatican Council, which met just before Darwin finished his *Descent of Man*, declared that Catholics are bound by faith to believe that we can attain the sure knowledge of God's existence by reason. It was the first ecumenical council since the Council of Trent, which ended in 1563, over three hundred years earlier. Much had happened in the meantime in philosophy to undermine both reason and faith. So the council fathers thought it necessary to give a boost of confidence to the power of reason to help on the road to faith.

The beginning of modern philosophy is usually traced back to René Descartes (1596–1650), whose new approach to the subject gave rise both to modern empiricism, exemplified in different ways by John Locke and David Hume, and to modern idealism, such as that of George Berkeley or Baruch Spinoza. Although these thinkers differed much among themselves, epistemology—the study of how we come to know things and what we can know—became a central concern to them all.

The centrality of epistemology is evident especially in the writings of Immanuel Kant, who was awakened from his "dogmatic slumbers" by Hume's skepticism and tried to develop a new epistemology that, in his view, allowed for real knowledge of many aspects of the world. But, for Kant, this certainty did not extend to metaphysics. Human reason, in Kant's estimation, was powerless to establish the existence of God.

Kant's thought inspired a new line of philosophers, the most famous being Georg Wilhelm Friedrich Hegel. Since it is impossible to do theology without adopting some philosophical approach, it was only to be expected that various Catholic theologians should have borrowed concepts from the ascendant philosophers of the day, which included Kant, Hegel, and their leading disciples. Throughout the nineteenth century, the Magisterium of the Church was kept busy evaluating these various philosophical approaches and putting most of them on the Index of Forbidden Books.

A purely defensive response was hardly sufficient, however. Recognizing this, Pope Leo XIII, in his Encyclical *Aeterni Patris* (1879), urged Catholic philosophers and theologians to return to the *philosophia perennis*, best exemplified in the work of Saint Thomas Aquinas, as the time-tested optimal way to understand revealed truths and evaluate the challenges that modernity was bringing to the faith through advances in scientific knowledge and major societal changes. But already in 1870, the Fathers of Vatican I thought it important to address some of the challenges to the faith arising from modern philosophy. Hence the dogmatic definition that human reason can attain to the knowledge of God, or at least the God of the philosophers. By this is meant God as eternal, all-powerful, all-knowing spirit, the ground of all being. This definition, it should be noted, does not deny that grace is necessary to come to belief in the God and Father of our Lord Jesus Christ.

The teaching is clearly biblical. The two classical texts in support of the assertion come from Saint Paul and from the book of Wisdom. In his letter to the Romans, Paul writes, "Ever since the creation of the world his invisible nature, namely, his eternal power and deity, has been clearly perceived in the things that have been made" (Romans 1:20). In the Old Testament, the Book of Wisdom tell us this:

> For all men who were ignorant of God were foolish by nature; and they were unable from the good things that are seen to know him who exists, nor did they recognize the craftsman while paying heed to his works; but they supposed that either fire or wind or swift air, or the circle of the stars, or turbulent water, or the luminaries of heaven were the gods that rule the world. If through delight in the beauty of these things men assumed them to be gods, let them know how much better than these is their Lord, for the author of beauty created them. And if men were amazed at their power and working, let them perceive from them how much more powerful is he who formed them. For from the greatness and beauty of created things comes a corresponding perception of their Creator. (Wisdom 13:1–5)

The dogmatic definition made it clear that the biblical texts were to be taken at their face value, that they were just as true and relevant in the nineteenth century as they were at the time they were written. The Church does not provide us with a conclusive argument to God's existence. So it has been said that the definition is just an assertion of the power of human reason to attain truth. But it seems that the Church understands the definition to mean more than that.

The *Catechism of the Catholic Church* teaches:

> Created in God's image and called to know and love him, the person who seeks God discovers certain ways of coming to know

him. These are also called proofs for the existence of God, not in the sense of proofs in the natural sciences, but rather in the sense of "converging and convincing arguments," which allow us to attain certainty about the truth. These "ways" of approaching God from creation have a twofold point of departure: the physical world, and the human person.[3]

Anyone familiar with the history of the natural sciences cannot help but smile at the suggestion that any science aside from mathematics provides incontrovertible proofs, but the meaning of the passage from the *Catechism* is clear: there are several arguments, any one of which may not by itself have sufficient power to demand assent, but which taken together converge to make God's existence a certainty.

From the Physical World

The more traditional ways of coming to know the existence of God are from the physical world. Most students of philosophy are aware of Saint Thomas's five ways, which all begin from some aspect of the world around us. Many today find the first four of these ways either problematic or uninspiring. The first way, which Thomas calls the most manifest, is based on "motion." It begins with the observation that nothing is moved except by another. So if we encounter something moving, say a rock, we surmise that it was thrown, and so moved, by a hand; the hand, in turn, was "moved" by the will, which was, in turn, "moved" by the desire to hit the pigeon, which was in turn... This chain cannot go on forever but must stop at an Unmoved Mover, "and this everyone understands to be God." In modernity, lots of ink was spilled in debates as to whether Newton's principle of inertia vitiates the proof. Clearly something more than locomotion is meant in the proof, so inertia is a red herring. But, just as clearly, the first way does not make everyone drop to his knees in worship.

The second way similarly argues from the fact that everything we encounter in the world needs a cause, to an Uncaused Cause. The third way argues from the contingency of being to a Necessary Being. The fourth way argues from the fact that there are grades of perfection in goodness, truth, nobility, and the like, to the conclusion that there must be a being that is highest in goodness, truth, and nobility, which must be the highest in being and the cause of all being.

If this were a book of Catholic philosophy, space would be given to make a full-throated defense of these four arguments. But it is clear that every one of these ways requires a fair bit of philosophical sophistication to plumb, and in any case, this is not primarily a book of philosophy. It's a book about scientific evidence—which brings us to the fifth way. This argument is different from the first four, as is evident from Thomas's formulation in the *Summa Theologiae*:

> The fifth way is taken from the governance of the world. We see that things which lack intelligence, such as natural bodies, act for an end, and this is evident from their acting always, or nearly always, in the same way, so as to obtain the best result. Hence it is plain that not fortuitously, but designedly, do they achieve their end. Now whatever lacks intelligence cannot move towards an end, unless it be directed by some being endowed with knowledge and intelligence; as the arrow is shot to its mark by the archer. Therefore some intelligent being exists by whom all natural things are directed to their end; and this being we call God.

As it stands, the connection between Intelligent Design (ID) and the Fifth Way (FW) is not immediately obvious. And, in fact, the proponents of intelligent design rightly note that ID doesn't get one all the way to a transcendent God. But there is a connection between ID and the FW.

Dominican priest Michael Chaberek has an excellent discussion of the relationship between the two in his book *Aquinas and Evolution*. If the ways of coming to God are preambles to faith, then, as he explains, ID can be considered a preamble to the preambles of faith. This is because ID gives examples of specified complexity to illustrate why, as Thomas puts it, "philosophers call every work of nature the work of intelligence."[4]

The FW, Chaberek says, is based on the logical inference that a Mind is necessary to create the laws of nature, which account for the laws of natural bodies. A more updated version might invoke the fine-tuning of the laws of physics, which enable the universe to exist in an orderly form and, more than this, support life. ID, as it is commonly understood in biological circles, finds structures and processes in nature that can now persist through the laws of nature but which could only, because of their specified complexity, have arisen through the agency of a mind. For example, the laws of physics permit the existence of a DNA molecule and support its exact replication. But there is nothing in the laws of physics that would have led the elements in some primordial soup to coalesce spontaneously into information-bearing molecules capable of self-replication. ID shows this to be the case and then runs with it, making the argument for intelligent design as the preferred alternative, as the only causally adequate, and therefore best, explanation.

Saint Thomas did not have readily available examples of specified complexity such as are revealed by modern molecular biology, but it is possible to find a text that strongly suggests that he would be in favor of ID. In the *Summa Contra Gentiles*, he writes:

> That which results from the action of an agent, but apart from the intention of the agent, is said to happen by chance or by luck. But we observe that what happens in the working of nature is either always, or mostly, for the better. Thus, in the plant world, leaves are arranged so as to protect the fruit, and among animals the bodily organs are disposed in such a way that the animal can be protected. So if it came about apart from the intention of the natural agent, it would be by chance or by luck. But this is impossible, for things which occur always or for the most part are neither chance nor fortuitous events.[5]

Thomas is giving a basic lesson on the detection of purpose. A true roulette wheel, for example, is designed to produce chance events; that is to say, each of the numbers should come up with more or less the same frequency. A roulette wheel that kept giving the same number consistently would clearly not work according to chance, but would be fixed by the unscrupulous casino owner. In Thomas's view, it is clear that organic structures result from the intention of an agent, because they occur regularly and for a reason. And only intelligent agents can have intentions.

At this point, one can engage in learned arguments as to how the intelligent agent causes the highly specified information-rich elements of life. One can find theistic evolutionists singing the praise of providence that can use "chance" to do its work. But the problem is that these same people tend to look upon chance as something that escapes God's foreknowledge, as will become apparent in Chapter 8. And it should be clear to the reader by now that there is not enough probabilistic potential in the universe for life to have assembled by chance. The grasping at the weak straw of infinite multiverses is a clear admission of that basic fact. As the physicist Bernard Carr said, "If you don't want God, you'd better have a multiverse."[6] Once chance is eliminated, we are left with intimations of God's personal involvement on the evolutionary path to the flora and fauna of today—or alternatively, of His particular intelligent fashioning of each such form.

The design inference touches something deep in the human psyche. The various authors of the Old Testament used the order of nature and the intricacies of creation to praise God. Psalm 19,[7] for example, begins, "The heavens are telling the glory of God; and the firmament proclaims his handiwork." Psalm 104 is a long exposition of all the wonders that God has created. A snippet suffices to illustrate: "The trees of the Lord are watered abundantly, the cedars of Lebanon which he planted. In them the birds build their nests; the stork has her home in the fir trees. The high mountains are for the wild goats; the rocks are a refuge for the badger" (Psalm 104:16–18). The listener is supposed to be enthralled by how everything fits together and to raise his voice in praise of the Creator.

And it is not only in the Book of Psalms where we find such examples. God silences Job by an overwhelming list of created beings that are clearly beyond human powers to create. Again a snippet from a much longer passage suffices here: "Is it by your wisdom that the hawk soars, and spreads his wings toward the south? Is it at your command that the eagle mounts up and makes his nest on high? On the rock he dwells and makes his home in the fastness of the rocky crag. Thence he spies out the prey; his eyes behold it afar off" (Job 39:26–29). And in the prophet Daniel we find the song of the three youths who were thrown into the fiery furnace: "Bless the Lord, you whales and all creatures that move in the waters, sing praise to him and highly exalt him for ever. Bless the Lord, all birds of the air, sing praise to him and highly exalt him for ever. Bless the Lord, all beasts and cattle, sing praise to him and highly exalt him for ever. Bless the Lord, you sons of men, sing praise to him and highly exalt him for ever" (Daniel 3:57–60).

These examples suffice to illustrate the lament found in the Book of Wisdom: "For all men who were ignorant of God were foolish by nature; and they were unable from the good things that are seen to know him who exists, nor did they recognize the craftsman while paying heed to his works" (Wisdom 13:1). The passage continues by suggesting that the created things are so wonderful and beautiful that they captured the full attention of those who investigated them, but since those investigators did not draw the logical conclusion, they were blameworthy: "Yet again, not even they are to be excused; for if they had the power to know so much that they could investigate the world, how did they fail to find sooner the Lord of these things?" (Wisdom 13:8–9).

In the New Testament, Christ continued this line of reasoning. In Matthew, we read:

> Therefore I tell you, do not be anxious about your life, what you shall eat or what you shall drink, nor about your body, what you shall put on. Is not life more than food, and the body more than clothing? Look at the birds of the air: they neither sow nor reap nor gather into barns, and yet your heavenly Father feeds them. Are you not of more value than they? And which of you by being anxious can add one cubit to his span of life? And why are you anxious about clothing? Consider the lilies of the field, how they grow; they neither toil nor spin; yet I tell you, even Solomon in all his glory was not arrayed like one of these. But if God so clothes the grass of the field, which today is alive and tomorrow is thrown into the oven, will he not much more clothe you, O men of little faith? (Matthew 6:25–30)

The biblical authors and Christ Himself were taking the wonders of creation as clear arguments for the existence of God and His providence. What else could Saint Paul have had in mind when he asserted that God could be known from creation? One cannot imagine Christ telling his hearers: all that is moved is moved by another, so unless there be an unmoved mover, no motion will be possible. Those on the shores of the Sea of Galilee would have been totally confused and would have thought Him mad; and modern philosophers would have been eager to point out problems. Yet the simple direct appeal to the intricacies and beauties of creation has power to convince both the simple and the wise.

One can find the same strategy in patristic writings for showing that God exists. John West cites the example of Theophilus (ca. 115–188): "Theophilus went on to list the functional regularities of nature from astronomy, the plant world, the diverse species of animals, and the ecosystem. His conclusion? Just 'as any person, when he sees a ship on the sea rigged and in sail, and making for the harbor, will no doubt infer that there is a pilot in her who is steering her; so we must perceive that God is the governor [pilot] of the whole universe.'" West also cites similar passages from Dionysius (200–265), the Bishop of Alexandria, and Lactantius (c. 240–320). "Such citations from the Fathers of the early church can be multiplied," writes West. "Early Christians clearly and repeatedly taught that nature provides convincing evidence of God's design."[8]

Skipping ahead to modernity, we encounter Immanuel Kant, who was moved by the argument for God's existence from design. He classified the first four of Saint Thomas's ways as cosmological, and dismissed them as invalid in light of the limitations of human reason as he conceived it. But as to the argument from design, which he called the "physico-theological" argument, he said: "This proof always deserves to be named with respect. It is the oldest, clearest, and most appropriate to common human reason. It enlivens the study of nature, just as it gets its existence from this study and through it receives ever renewed force."[9] Admittedly, Kant thought that it did not quite succeed in proving God's existence, because it depends on the validity of the cosmological argument to establish that the designer must be God. And Kant maintained that the validity of the cosmological argument required the support of the ontological argument, which he rejected. Nevertheless, as Roger Scruton comments, despite these reservations toward the physico-theological argument to God, Kant "felt it to be the expression of a true presentiment."[10]

It is not essential that the design argument be as watertight as a proof in geometry. The point is that it is powerful. And it is based on a logic that the average person can follow. It stands as one of the "converging and convincing arguments" that allow us to attain the certain knowledge that God exists. If ID is dismissed, it is difficult to see what other way remains open as an argument for God's existence from the physical world, certainly none that would be in continuity with Scripture and Tradition.

Saint John Paul II gave a talk in 1985 in which he warned of the dangers of rejecting the design argument. "The evolution of living things, of which science seeks to determine the stages and to discern the mechanism, presents an internal finality, which arouses admiration," said the Pope. This finality "directs beings in a direction for which they are not responsible or in charge" and thus "obliges one to suppose a Mind, which is its inventor, its creator." He continues:

> To all these indications of the existence of God the Creator, some oppose the power of chance or of the proper mechanisms of matter. To speak of chance for a universe which presents such a complex organization in its elements and such marvelous finality in its life would be equivalent to giving up the search for an explanation of the world as it appears to us. In fact, this would be equivalent to admitting effects without a cause. It would be to abdicate human intelligence, which would thus refuse to think and to seek a solution for its problems.[11]

It is not surprising that atheists should want to reject ID, but it is disconcerting that many Catholics do. Their motivations might be various, ranging from the desire to keep their scientific positions or prestige—human weakness—to a genuine worry that future progress in science might reveal that the basic laws of physics can account for the information packed into living beings—the "god-of-the gaps" worry. Or they might want to protect the creating intelligence from responsibility for designing what appears manifestly evil, such as the predatory mating tactics of the black widow or praying mantis. Or they might think that the argument is too crude, for they see it working by analogy, and when terms are applied to God by analogy, there is more unlike than like in them.

Some of these concerns will be addressed at length in subsequent chapters or have already been discussed in the preceding pages. Briefly, and to reinforce by repetition, Catholics are obliged to believe in some limitations (so-called "gaps") in the causal powers of purely natural processes to produce what we find around us, as we saw in Chapter 3. We know by faith, for example, that the human soul does not arise out of matter, so neuroscience will never explain how a particular configuration of matter can attain self-consciousness. (And here I do not mean to imply that there are not also empirical/scientific reasons for concluding this. I am convinced there are.) As to evil, if it really is such, which in a fallen world it may be, it does not invalidate the need for an intelligent designer. The death chambers at Auschwitz are unambiguously evil, and yet they did not arise by chance. Finally, a bit of philosophical humility might be in order for those who want to protect God from what they regard as false analogical attributions. "Truly, I say to you, unless you turn and become like children, you will never enter the kingdom of heaven" (Matthew 18:3).

Philosophically sophisticated Catholics, who know that they must in principle believe that natural reason can arrive at the God of the philosophers, are often willing to see a design argument in the fine-tuning of the laws of physics but not in the intricacies of the biological world. They see the latter as an aggressive evangelical Protestant creationism, whereas Catholics, in their opinion, have grown beyond that and have learned to see God in the big picture without linking Him directly to the wonders of biology. But in order to do that, and live in a world filled with life, they have to support the empirically unsubstantiated claim that natural selection and random genetic mutations can take care of all of life once the rules of physics and other initial conditions are properly set up. Their resultant idea of God tends to be deistic: God created the laws and now He is no longer involved, or at least very little involved. It is far from the picture of God presented by Jesus in the Gospels: "Are not two sparrows sold for a penny? And not one of them will fall to the ground without your Father's will. But even the hairs of your head are all numbered" (Matthew 10:29–30). Knowingly or not, such Catholics are supporting efforts to keep the divine foot out of the door.

A 2019 debate over intelligent design between Michael Behe and Stacy Trasancos illustrates the reticence some Catholics feel toward ID.[12] Behe, as noted, is a professor of biology at Lehigh University. Trasancos holds a doctorate in chemistry and is a convert to Catholicism. She very much wants to pass on the faith to others. She says that she tried to be more sympathetic to ID but found that it was hindering her efforts at evangelization and did not square with her understanding of the relationship between faith and science.

It is not that she is against seeing the world as intelligently designed. But she believes it to be intelligently designed because her faith teaches her that God created it *ex nihilo*. Her mantra is to "go bigger." She sees design in the elements in the periodic table and in the sand on the seashore. This "bigger" design, it is supposed, led inevitably to the emergence of biological life and its evolutionary development. So, in a sense, everything is designed.

Behe, of course, agrees that God designed it all. But he also sees design within design. The chemical elements are already a wonder of design. But the living machines that are built out of them require another level of design. We are all familiar with LEGOs. Clearly the pieces are the product of intelligence. We can see this even if they are scattered all over the table, in no particular pattern. But we see another level of design if we encounter a fully constructed LEGO model of the Vatican. And this second level of complexity, clearly evident in biology, is empirically detectable, for it is statistically impossible that it could be the product of chance and the regularities of nature alone.

The noun "chance" and its adjective cousin "random" are tricky words, because we apply them to events that are unpredictable due to our not having anywhere close to a fully detailed causal account that would allow us to securely predict their outcome. But one could also say that while many events are random from our point of view, they are not so to God, since God knows everything. No doubt this is true. But when Darwinists talk about random mutations, or chance + natural selection, they do not mean *apparently random/chance-like but in fact carefully orchestrated by an overriding intelligence.* They mean it in the ordinary sense of a happenstance occurrence not under the control of any overriding intelligence. And yet we have already seen that the odds of stumbling upon a functional protein are often less than one chance in 10^{77}, and that even the simplest life requires far more than a single functional protein to survive and reproduce, so a random process cannot account for what we see before us, with or without natural selection waiting in the wings to lend a hand. The chemical elements, themselves carefully designed to make life possible, do not have it in themselves to come together to form living beings. As with the LEGO blocks, an additional visitation of intelligence is required, beyond what is inherently in them.

Trasancos is concerned not to discourage potential converts to Christ by demanding that they reject Darwinism. Rather, she wants to teach them that faith and science are compatible. Her goal is laudable. But it is not a reason to disparage ID. In the interview, she agrees that it is legitimate to reject bad philosophy. Well, that is exactly what Darwinism is—an uncompromising materialism utterly resistant to contrary empirical evidence. As such, it is incompatible with both good science and Christianity. Christians need not deny evolution in every sense. Behe himself insists that universal common descent is perfectly tenable, as long as one acknowledges that matter by itself could not be responsible for the myriad life-forms that have inhabited our planet.

Trasancos wants to keep science and theology separate. It is not that she thinks that her scientific work is in some ways incompatible with her Christian faith. But she rightly thinks that it is possible to do science, to explore the natural world, without explicitly invoking her faith. She acknowledges the debt that science owes to Christianity and has written a book to promote this idea: *Science Was Born of Christianity*, which presents the perspective of Fr. Stanley Jaki on this subject. But she sees no need to keep stressing design in scientific publications. Although she believes that everything is designed, she does not write at the end of her papers in learned journals of chemistry that whatever she just presented is a wonderful example of God's providence. So it disturbs her to see affirmations of design in the works of Behe and other scientists. She thinks that by pointing out that something like the flagellum was designed, one detracts from the design that is found absolutely everywhere in nature.

There is something that can be said in favor of such an approach. There is no "one size fits all" approach to evangelization, so I would not want to insist that she change her style. But the other tactic also has something to be said for it. Papers in biology are peppered with references to evolution, just to evangelize for the Darwinian account, even when the study has made no effort to explain how evolution might have actually contributed to the phenomenon in question. The same is true for museum displays and other public presentations of science. So although there may not be a need to bring up the question of design explicitly in journals of chemistry, there is a need to invoke ID in biology as a countermeasure to misleading Darwinist propaganda.

Finally, Trasancos says that she sees design everywhere because she starts from faith. Faith seeking understanding must be the major route for Christians, especially when it comes to understanding what is clearly beyond the reach of unaided human reason. But where does faith come from? Trasancos does not like ID because she sees it as needing miracles, and what is described as miraculous today, might be explained by science tomorrow—the god-of-the-gaps worry. Ultimately, however, the Christian faith does depend on miracles. Why should we believe in Christ? Jesus himself tells us: "even though you do not believe me, believe the works, that you may know and understand that the Father is in me and I am in the Father" (John 10:38). It was the miracles that caught the attention of Jesus's contemporaries. And he sealed his teaching with the ultimate miracle of the resurrection. Without it, no one today would have heard of him.

What does the theory of intelligent design demand? Less than you might suppose. Some leading proponents of intelligent design see evidence against universal common descent as decisive, but Michael Behe, one of design theory's leading voices, remains comfortable with universal common descent and does not call for miracles in any conventional sense. He is open to the idea that subtly guided quantum processes could be the means by which the designer endows biology with new information.[13] So, detecting design in biology, and concluding that some or another feature of an organism is the result of design, is not identical with invoking a miracle traditionally understood—though any Catholic should be open to the possibility of such miracles.

The whole ID enterprise is sometimes dismissed as just a resurrection of William Paley's famous watchmaker argument—and it is conventional wisdom in the academy that Paley's watchmaker argument has been refuted. Actually, only those who believe, against all empirical evidence, that the Darwinian mechanism alone can produce life in all its forms can confidently ridicule Paley. Yes, Paley was overly optimistic about how much biology can reveal of its creator—God's benevolence toward all creatures, for example. But Paley's argument stands, at least as a preamble to one of the preambles of faith. No one in the contemporary ID community claims that ID by itself gets one to the existence of the Christian God. But it does get one well down the path.

The Ways to God from the Human Person

The *Catechism* says, "With his openness to truth and beauty, his sense of moral goodness, his freedom and the voice of his conscience, with his longings for the infinite and for happiness, man questions himself about God's existence. In all this he discerns signs of his spiritual soul. The soul, the 'seed of eternity we bear in ourselves, irreducible to the merely material,' can have its origin only in God."[14]

There are three ways in which the acceptance of the grand claims of Darwinism negate the approach to God from the human person. The first is the insistence that all the powers traditionally ascribed to a spiritual, i.e., immaterial, soul are just epiphenomena of matter. It is the position presently held by most academics in the West. But it is the very thing Saint John Paul II condemned in his allocution to the Papal Academy of Science on evolution: "The theories of evolution which, because of the philosophies which inspire them, regard the spirit either as emerging from the forces of living matter, or as a simple epiphenomenon of that matter, are incompatible with the truth about man."[15]

But it is not just the assertion that man is fully explainable by physics and chemistry that undermines the argument for God's existence from the nature of the human person. There is a more specific roadblock constructed from the conclusions of evolutionary sociology or evolutionary psychology.

The traditional argument to God from the human person is that each of us finds within himself or herself a natural law, mediated to us by our conscience. We know, for example, that killing the innocent for the fun of it is wrong. Even something less obvious, such as lying, pricks the conscience. Young children tend to look at the ground when lying to their parents. And lie detectors pick up on our bodies' natural rebellion against lying. We each encounter the natural law within us. But we know that we did not put it there. So it must have been given to us by a Law-Giver—that is, God.

Apparently not, according to evolutionary sociologists and evolutionary psychologists. The most obvious cases will have occurred to any astute reader of Darwin. We are lustful because natural selection favors those who copulate often. We are violent because natural selection eliminates those who are slow to strike. We are greedy because only those who hoarded enough food to survive the winter or the drought went on to reproduce. These are obvious implications of the theory. But the evolutionary sociology/psychology industry keeps churning out ever more creative just-so stories of this sort in academic papers. These are then popularized on the internet and in weekend editions of newspapers.

Here is a small sample of what one can find on the internet: "Why They Stray: The Evolutionary Advantages of Infidelity"[16]; "Depression's Evolutionary Roots"[17]; "Why Cry? Evolutionary Biologists Show Crying Can Strengthen Relationships"[18]; and "Understanding Lies Based on Evolutionary Psychology: A Critical Review."[19] It is true that sometimes the authors disagree with one another. But they remain stuck in the paradigm that every human behavior can and must be explained in terms of Darwinian tactics. The idea that God created us with a conscience to help us reach our potential as his sons and daughters is off the map for them.

Fortunately it is easy to dismiss the grand claims of evolutionary psychology, because evolutionary psychology provides a basis for its own refutation. The commonly held evolutionary epistemology says that our thought patterns are the products of a blind process that preserves survival-enhancing behaviors and features. On this view, the reason we believe that we can attain general truths about the world from several instances—the problem of induction and abstraction—is that holding such a belief was advantageous to our ancestors in navigating the world. Those who spent too long thinking about whether we could know anything were eliminated from the gene pool. "He who hesitates is lost," as the saying goes. What follows from this is that Darwinism itself gives us reason to conclude that even our firmest convictions that we know this or that about the world may be illusions fobbed off on us by evolution. This would include, of course, an evolutionist's conviction that he knows evolutionary theory to be true.

Jerry Fodor and Massimo Piattelli-Palmarin, in their *What Darwin Got Wrong*, give several examples of prominent thinkers who turned to Darwin to develop an epistemology. Among these is the philosopher Willard Van Orman Quine, who opined, albeit tentatively, "Why does our innate subjective spacing of qualities accord so well with the functionally relevant grouping in nature as to make our inductions come out right? Why should our subjective spacing of qualities have a special purchase on nature and a lien on the future? There is some encouragement in Darwin."[20] If Quine was somewhat hesitant, Karl Popper and Donald Campbell, who coined the term "evolutionary epistemology," were positively exuberant about the Darwinian solution.

Fodor and Piattelli-Palmarin quote Campbell's all-encompassing vision: "The natural selection paradigm for such knowledge increments can be generalized to other epistemic activities such as learning, thought, and science."[21] This natural process, for example, is said to have given rise to Kant's categories of perception and thought. That's not to say that Kant was influenced by Darwin (who came after him), but that moderns, such as Campbell, credit natural selection with producing these aspects of human cognition and reasoning. As for Popper, although he said that Darwinism was not a testable theory but a metaphysical research project, he nevertheless wrote, "Yet, the theory is invaluable. I do not see how, without it, our knowledge could have grown as it has since Darwin."[22]

Fodor, an atheist, is a philosopher of mind. He wrote his anti-Darwinist book precisely because he thinks that Darwinism has made nonsense of serious work in his field. When evolutionary psychologists and epistemologists protested that "no one is that kind of Darwinist anymore," he provided much evidence that various mainstream contemporary thinkers were precisely that kind of Darwinist.

The third danger that Darwinism poses to our understanding of the human person is that it denies an unchanging human nature. Darwinism is a form of nominalism, because each individual is human only in the sense that he more or less looks like others who are called human, and he can mate with individuals of the opposite sex and produce viable offspring. Although Darwin's magnum opus is entitled *The Origin of Species*, he was aware that he was destroying the very notion of species as a reality in nature: "In short, we shall have to treat species in the same manner as those naturalists treat genera, who admit that genera are merely artificial combinations made for convenience. This may not be a cheering prospect; but we shall at least be freed from the vain search for the undiscovered and undiscoverable essence of the term species."[23]

Many of Darwin's contemporaries were disturbed by this idea, for they were "unrepentant realists about species."[24] One needs to be a realist, if not about species as a Linnaean category, then at least about species as natural kinds, if one is going to maintain that species each come with a natural law. Some follow this law without being conscious of it, because they are not sentient. It is a natural law of sunflowers to track the sun or of fish to need water to live. But for humans, who can choose their actions freely (within, of course, the limitations of physics), the natural law entails moral imperatives. But in the Darwinian scheme, the concept of a natural moral law for human nature makes no sense; there is no one single natural law that applies to everyone who is biologically classified as human that can point us to a Lawgiver—God.

According to the Darwinian framework, what we discern within ourselves as law is nothing but an admixture of nature and nurture, and it can vary among contemporary individuals and change throughout history. It ceases to be an authoritative basis from which to ground moral reasoning. If human nature is plastic, as Darwinism holds, then our moral discourse is ungrounded. The way to God from the human person is blocked. And God, in turn, cannot be the foundation of morality. (He is, of course, more than that; but He is not less.)

If an atheist such as Jerry Fodor should be offended by the legacy of Darwinism because it dissolves the power of the human soul to attain truth, there is all the more reason why Catholics should reject it. There is no scientific reason to accept this universal acid of skepticism as we walk the paths towards God.

5. Creation Groans

For the creation waits with eager longing for the revealing of the sons of God; for the creation was subjected to futility, not of its own will but by the will of him who subjected it in hope; because the creation itself will be set free from its bondage to decay and obtain the glorious liberty of the children of God. We know that the whole creation has been groaning in travail together until now.
—Romans 8:19–22

The above passage from Saint Paul's letter to the Romans sheds some light on a great mystery—the cruelties and imperfections of nature as we now encounter it. The cruelties are evident to anyone with a pair of eyes. Darwin, in a letter to Asa Gray, cited the parasitic wasp as a reason not to believe that God had created the various species. This insect lays its eggs in a live caterpillar, whose young then consume their host from the inside, before eating their way out. In the same passage, Darwin also mentioned the cruelties of cats with mice.[1]

The imperfections take a little bit more investigation. Darwinists point to instances of what they assure us are poor design in biology as evidence that the hit-or-miss process of evolution, rather than the wise planning of an all-knowing God, is the cause of life's diversity. They point out, for example, the "reversed wiring" of the human eye or the "convoluted position" of the laryngeal nerve in a giraffe, design "choices" they say no self-respecting engineer would have let pass. These are often used as evidence that ID must be wrong and hence Darwinism, the only "viable" alternative, must be right.

Figure 5.1.
Parasitic wasp cocoons, *Cotesia congregata*, on the *Manduca sexta* caterpillar. Cocoon size is about 3 mm.

The importance of addressing these cruelties and imperfections can hardly be overstated. Cornelius Hunter has argued that the most persuasive arguments for Darwinism, which go back to Darwin himself, are based on popular ideas of God.[2] The early nineteenth century had inherited a tradition of rational arguments for the existence of God. This might, at first, sound Catholic, exactly in line with the Scriptures, but the program had overshot its mark and opened itself up to refutation.

Saint Paul speaks of the possibility of coming to know the "eternal power and deity" of God. But William Paley, who supplies us with the most famous example of natural theology in England, thought he could go further and also demonstrate the benevolence of God from nature. He was aware, of course, that carnage happens, but, among other arguments, he appealed to the fact that in most cases animals experienced a quick death. Nature was, in his estimation, by and large very pleasant, reflecting the goodness of the Creator: "It is a happy world after all."³

Darwin's contemporaries thought that they knew precisely how God should act if indeed he were wise and good. And when they looked closely at nature, they were appalled—carnage, slow death, inefficiencies everywhere. This line of argument for evolution, which proved persuasive to many, was not based on empirical observation of natural selection in action or a continuum of fossils. Rather it was a negative argument against God based on conviction as to what God *should* have done. To Darwin and his contemporaries, the only obvious alternative to evolution by natural selection was special creation by God. And the carnage, waste, and imperfections of nature functioned as arguments against this alternative. These arguments proved rhetorically powerful and helped evolution win the day. Darwinism, according to Hunter, was a kind of theodicy—a justification of God. God was justified against the charge of putting evil in nature by being removed from the scene.

This rationalistic account of God is still the prevalent one today. Hardcore atheists such as Dawkins depend on it to carry on the same kind of propaganda Huxley and Haeckel practiced in the nineteenth century.⁴ It is the fundamental presupposition behind all the arguments in favor of Darwinian evolution based on cruelties and imperfections.

But the God revealed in the Bible is much greater and more mysterious than his pathetic and tidy rationalist counterpart. He created *ex nihilo* and He keeps creation in being and guides it by his Providence. He caused a great Flood. He punished his people and their enemies—witness the plagues in Egypt and the fiery serpents that were sent to kill the complainers during the Exodus. Ultimately this God entered the messy fallen world in which we live and bled to death on a cross. He is a Mystery. And the world He created is a mystery.

Thus, the strongest argument for Darwinism is based on bad, non-biblical theology. And now the same bad theology is being used to discredit the preamble to the preamble to the faith: intelligent design. The objections to ID are philosophical. So, for the most part, the responses to them are philosophical, although some biological details will help to illustrate the argument. I cannot cover all the usual examples. But I hope that the few examples I provide will help the reader not to be disturbed when future examples of "cruelties" and "imperfections" are presented as arguments in favor of Darwinism or of any other supposedly mindless creation account.

Cruelties

Insects provide the most obvious examples of behavior that makes us step back in horror. They have appendages that are clearly designed to maim, decapitate, poison, and destroy. Dragonflies, scorpions, and praying mantises are the inspiration for monsters in horror movies. The predatory mating antics of some of these insects are especially revolting.

Look at the praying mantis. The much smaller male praying mantis, attracted by the pheromones of the larger female, has to jump stealthily onto her back to start the copulation. Those who are not quick enough end up being food for her: a pre-coital snack. But even the lucky one who manages to latch on soon gets decapitated and parts of him eaten. Nevertheless, the decapitated male keeps pumping sperm into the female while she is eating him. There is a secondary nerve center—a brain—in his abdomen, which ensures that seed will be passed on to create a new generation of these monsters.[5]

The aptly named black widow also kills and eats her much smaller mate after copulation. For his part, the male leaves behind an appendage in her to block any future male from depositing his sperm.[6]

Most of us have encountered ant wars in person, in a backyard or in a park. If not, there are many videos on the internet showing gruesome details that escape the unaided eye. There is the warfare of red ants versus black ants. There is the warfare of black ants versus black ants. And the warfare is brutal. One colony often wipes out another completely. And they seem to be doing it for reasons that make sense to us: to kill a rival queen before her colony kills theirs or to steal their resources.

Darwin was especially bothered by ants because he came from a family that was opposed to slavery. Yet he witnessed before his eyes the raids of slave-making red ants on a colony of black ants to enslave them.[7] This presented a great quandary for Darwin, because it supposedly showed him that slavery was "natural." In his heart, Darwin had sympathy for his fellow man, although with his head he insisted that inferior races, and by this he meant everyone but Caucasians, must eventually be wiped out in the evolutionary struggle for survival.[8]

The cruelties of nature, moreover, are not limited to the insect world. Most people would be happy to take a can of insecticide to whatever six- or eight-legged critters they encounter. But we tend to have more sympathy for various other kinds of animals. *National Geographic* seldom shows the actual moments when lions kill zebras. They show the chase; and they show the feast afterwards when the carcass no longer looks like much of anything. The neck-breaking bites and take-downs on the Serengeti are usually skipped over. Nor is it just the iconically ferocious beasts who engage in savage destruction. The cute house sparrow, abundant in city parks and gardens, is known to take over bluebird nests, destroy the eggs, kill the adult birds, and use their feathers as additional nest lining.

Animals other than insects also have strange sexual anatomies and behaviors. Ducks will serve to illustrate this point. Most birds—in fact 97 percent—do not have penises; they have cloaca, which are openings to their testes (male) and ovaries (female). Mating is by "cloacal kiss." Among the exceptions are various species of ducks. And they have the most bizarre penises and vaginas.

There is a duck in Argentina whose penis, when erect, is four inches longer than its body. Drakes usually keep this appendage housed inside their body, but it can be shot out to full length in about one third of a second. Stranger still is that the duck penis is shaped like a corkscrew and often equipped with ridges and backward-pointing spines. The Muscovy duck, for example, has between six and ten twists over its 20 cm length, which is about a third of its body length. The female vaginas are just as strange. They have dead ends and canals that are also twisted in a corkscrew shape—opposite to the spiral of the penis.

Why this bizarre female anatomy? It is clear that drakes rape. So researchers think that the elaborate vagina of the female duck allows her some protection because she can guide the penis of the rapist into the dead ends whereas she can allow the drake of her choice to deposit his sperm to fertilize her eggs. Biologists estimate that only 3 percent of the rapes lead to conception.[9] But that is still quite a number of ducklings produced in this manner, because "forced copulations are pervasively common in many species of ducks." These attacks are "violent, ugly, dangerous and even deadly.... Groups of males travel together and attack a single female in a form of gang rape."[10]

Other instances of revolting behavior involve dolphins who kill porpoises for no apparent reason and seals that rape penguins. Sea otters are especially "evil." They have been described as "disease-ridden, murderous, necrophilic aqua-weasels" for their penchant for killing baby seals and copulating for hours with their dead bodies.[11]

Many male animals seek out and kill the young of their species.[12] The explanation given works from an evolutionary perspective. A male encountering a female with suckling young will kill them so that she will return to estrus more quickly, mate with him, and ensure that it is his genes that will be propagated to future generations. This death toll can be quite high. It may be the leading cause of infant mortality in Chacma baboons.[13] Maternal infanticide is much rarer, but it too occurs.[14] Giant pandas often bear twins, but Mom tends to pick one and abandon the other to death by starvation.[15]

It is clear that sex and death are everywhere in nature. This is most clearly seen in the various changes in animal anatomy and behavior in mating season: plumage, antlers, coloration in fishes such as salmon, and dances and fights, sometimes to the death, for mates among cougars, bison, kangaroos, and many other animals. It is easy to think that Darwin must be onto something important.

Such a reaction is exemplified by David Hull. In a review of Phillip Johnson's *Darwin on Trial*, he wrote, "What kind of God can one infer from the sort of phenomena epitomized by the species on

Darwin's Galápagos Islands? The evolutionary process is rife with happenstance, contingency, incredible waste, death, pain and horror.... The God of the Galápagos is careless, wasteful, indifferent, almost diabolical. He is certainly not the sort of God to whom anyone would be inclined to pray."[16]

Hull is far from alone in his reaction. The existence of suffering and evil in the world is the most powerful argument against the existence of God. But it is not conclusive, especially for those who believe that God does not cause evil but only permits it in order to bring about a greater good through it. The Christian has the further consolation of knowing that, in Christ, God has experienced human suffering and triumphed over death. Although reactions such as Hull's might serve as excuses for atheism, it is important to disentangle the emotions from reason. A gut reaction against the cruelties of nature is not an argument against design. The empirical evidence of specified complexity in life demands a designer. You cannot overcome impossible odds with a queasy stomach.

It is, of course, legitimate to wonder about the myriads of cruel and stomach-turning designs in nature. They are just as real as the lovely and majestic designs recruited to sing God's glory. It is certainly fair to note that one cannot imagine a song of praise such as, "All ye black widows who devour their mates, Bless the Lord; All ye maggots that eat dead flesh; Bless the Lord; All ye mallards who gang rape, Bless the Lord; All ye viruses that cause plagues, Praise and exalt Him above all forever." Such images would never make it onto the inspirational posters sold in religious bookstores. But they are just as real a part of our world as the soaring eagles and majestic sunsets that adorn many parish halls and offices.

In pondering this mystery, we must remember that we live in a fallen world. The *Catechism* insists that original sin is "an essential truth of the faith." Adam and Eve were created immortal and experienced a great harmony of body and soul. Our first parents were naked and not ashamed. The animals and plants were subject to them. But then one of the fallen angels was given permission to test them. Why? Perhaps to give an opportunity to man and woman, such lowly creatures, to crush the devil and his pride. Alas, it was not to be. The opportunity to choose freely comes with the possibility of choosing wrongly; and Adam and Eve abused their freedom.

The *Catechism* describes the awful consequences of Adam and Eve's rebellion:

> The harmony in which they had found themselves, thanks to original justice, is now destroyed: the control of the soul's spiritual faculties over the body is shattered; the union of man and woman becomes subject to tensions, their relations henceforth marked by lust and domination. Harmony with creation is broken: visible creation has become alien and hostile to man. Because of man, creation is now subject "to its bondage to decay." Finally, the consequence explicitly foretold for this disobedience will come true: man will "return to the ground," for out of it he was taken. Death makes its entrance into human history.[17]

Darwinism is diametrically opposed to this essential Catholic teaching, for it sees death as an essential element to the formation of species, including man. Death becomes the creator. The struggle for survival is credited with shaping us and giving us the very characteristics that are ours as a result of original sin—lust, violence, greed. It is easy to see how these might be favored by natural selection.

The Catholic account of the creation and fall is a two-step process. We are raised by God to a pedestal and then, in Adam and Eve, we fall into the mire. Darwinism is a one-step process. All those who favor simplicity—Occam's razor—are naturally drawn to prefer Darwin's account. The simplicity, of course, is a chimera in the case of Darwin, because the empirical evidence is against him. Indeed, one could cogently argue that not even Occam's razor favors Darwinism in this case, since a crucial caveat of Occam's razor is to prefer the simpler solution "all other things being *equal*." And as we have seen, the Darwinian account is *unequal* to the task of explaining the origin of specified complexity, and it does a poor job of accounting for the reality of love, courage, and nobility, and of our sense that goodness is truly good and evil is indeed truly evil and not just a delusion fobbed off on us by our evolutionary past.

The *Catechism* teaching is based on biblical revelation. The book of Wisdom says, "God did not make death, and he does not delight in the death of the living.... It was through the devil's envy that death entered the world" (Wisdom 1:13 and 2:24). One suspects that this pertains only to human death, for the fossil record shows dinosaurs who looked well-equipped to kill and eat meat. Nevertheless, predation is something that another biblical text presents as symptomatic of the world in its present fallen state, at least implicitly. The prophet Isaiah paints an image of the world to come:

> The wolf shall dwell with the lamb, and the leopard shall lie down with the kid, and the calf and the lion and the fatling together, and a little child shall lead them. The cow and the bear shall feed; their young shall lie down together; and the lion shall eat straw like the ox. The sucking child shall play over the hole of the asp, and the weaned child shall put his hand on the adder's den. They shall not hurt or destroy in all my holy mountain; for the earth shall be full of the knowledge of the Lord as the waters cover the sea (Isaiah 11:6–9).

It may seem like an affront to the lion, with its razor-sharp teeth and retractable claws, to be reduced to eating straw. So one might still wonder why such a killing machine exists at all. And what about plague-causing viruses? And rapist drakes and otters?

A possible answer might run like this. The visible creation serves to reveal spiritual realities. A primary spiritual reality is the battle between the good angels and those who likewise were created good but chose not to serve God—the devil and his cohorts. Scripture in several places speaks of Michael and his angels leading the forces of God in battle against spiritual forces of evil.[18] This allegorical dimension of the creation would, on this view, have been kept at bay from Adam and Eve in the Garden of Eden, but after the fall, these protections were withdrawn, in part but not wholly by the couple being cast from the Garden because of their sin.

Much would now change for them. Women would find giving birth painful: "I will greatly multiply your pain in childbearing; in pain you shall bring forth children" (Genesis 3:16); and as for Adam: "cursed is the ground because of you; in toil you shall eat of it all the days of your life; thorns and thistles it shall bring forth to you" (Genesis 3:17b–18). Man could no longer stroll through paradise picking delicious ripe fruit to eat without a worry in the world. Outside the Garden, he found nature turned against him. He had to sweat to eke out a living. And various dangers that God had previously protected him from now found their way to him and his descendants: disease and predation, poisonous snakes and spiders, poisonous plants, hurricanes, tsunamis, wildfires—and murder by other human beings.

After the fall, the ant wars were no longer just an allegory of the spiritual warfare in the heavens; they became an image of what man now did to man. And perhaps the drakes and praying mantises also reveal some ugly truths about human sexual behavior. If nature is sacramental—a visible sign of God's glory—it can also be a visible sign of the present battle.

C. S. Lewis further speculated that such features of the natural world before the fall of man may not have been directly created by God but rather allowed:

> It is impossible at this point not to remember a certain sacred story which, though never included in the creeds, has been widely believed in the Church and seems to be implied in several Dominical, Pauline, and Johannine utterances—I mean the story that man was not the first creature to rebel against the Creator, but that some older, mightier being long since became apostate and is now emperor of darkness and (significantly) the Lord of this world....
>
> It seems to me, therefore, a reasonable supposition, that some mighty created power had already been at work for ill on the material universe, or the solar system, or, at least, the planet Earth, before ever man came on the scene; and that when man fell, someone had tempted him... If there is such a power, as I myself believe, it may well have corrupted the animal creation before man appeared.[19]

If we are able to modify life genetically, perhaps even produce lethal viruses, then surely demons, had God allowed them, could have done the same. Perhaps Adam and Eve were protected by a special grace from the dangers of this world up until the time that they rebelled. One might argue that even now animals often, though not invariably, retain some respect for humans: a vestige of the original plan for creation.

Many questions might arise at this point. If Adam and Eve had not sinned, would their offspring not have filled up the earth in very little time, especially since none of them would die? What are we to think about the wonderful processes that keep the earth fit for human life but which are potentially lethal to human life? Hugh Ross, for example, mentions tectonic activity, which "generates continents, which together with the oceans, recycle nutrients and steadily remove potentially destructive greenhouse gases from the atmosphere."[20] Yes, wonderful, but it also causes earthquakes and tsunamis. Would God have warned Adam and Eve and their righteous descendants in plenty of time before the arrival of a tsunami so that they could climb to a height and applaud the wave?

These are serious questions; and Christian theology has not provided detailed answers, even if plausible scenarios can be proffered. The general response is that God is not bound by the physical limitations of the world as we encounter it. It is a matter of faith that Jesus and Mary are in heaven in body and soul. Yet no one seriously expects to find them using a rocket ship. That means that there must be something like a parallel universe (though on a higher plane of existence) which is similar enough to ours so as to make sense of a bodily presence, but which is totally inaccessible to our scientific exploration. Moreover, the Church teaches that there will be an end to the universe as we now know it and as science describes it: "Then I saw a new heaven and a new earth; for the first heaven and the first earth had passed away, and the sea was no more" (Revelation 21:1).

What of the old cosmos? It's too simple to say that it will simply be dispensed with, for Jesus promises to make "all things new."[21] In the end, the whole universe will be transformed because of events that took place on this "pale blue dot," as Carl Sagan famously dismissed Earth.[22] In a similar vein Steven Weinberg wrote, "This present universe has evolved from an unspeakably unfamiliar early condition, and faces a future extinction of endless cold or intolerable heat. The more the universe seems comprehensible, the more it also seems pointless."[23]

The Christian vision of past, present, and future is a direct affront to this atheistic outlook.

One can almost hear the Darwinist reaction to Christian eschatology. It echoes Festus's interjection when Saint Paul spoke of the resurrection of Christ: "Paul, you are mad; your great learning is turning you mad" (Acts 26:24). Paul, of course, knew that no truly dead person (before Jesus) had ever risen to a new and perfect life. But that was the whole point. Christ did rise from the dead, which gave credibility to everything that He had said and done. The constant cycle of birth and death, growth and decay, which seemed to be self-enclosed and governed by necessity was, at least in the case of humans, broken. And if Christ could break that cycle—delivering on his promise—then there is every reason to believe that He is God and can indeed make all things new. Yes, the Christian hope is bold. But it is well grounded.

Imperfections

Alongside the cruelties of nature, its imperfections provide complementary arguments against ID. The vertebrate eye is a case in point. The fascination with the design of the eye goes back a long way. Paley used it as a paragon of design. Darwin used his imagination to overcome the difficulty of explaining its perfection. His modern followers, in contrast, never tire of drawing our attention to a particular "imperfection."

The alleged poor design is that the light receptors do not face the iris but rather the retina. What they are sensing is the light that lands on the retina. The "wires" (nerves) that carry the signal to the brain are then bundled together and guided through a hole in the retina. This causes a blind spot. Admittedly, it requires some skill to even become aware of this blind spot, so tiny is it. And all acknowledge that it does not harm the vision of anyone with two eyes. But the detractors of ID revel in pointing out that this "backward" wiring is messy and inelegant, hardly what we could expect from an all-wise designer.

Jonathan Wells provides a sample list of Darwinists pressing this argument. Dawkins (1986): "It is the principle of the thing that would offend any tidy minded engineer"; George Williams (1992): "There would be no blind spot if the vertebrate eye were intelligently designed"; Kenneth Miller (1994): "No one, for example, would suggest that the neural wiring connections should be placed on the side that faces the light, rather than on the side away from it. Incredibly, this is exactly how the human retina is constructed"; Douglas Futuyma (2005): "no intelligent engineer would be expected to design" the "functionally nonsensical arrangement." Wells's list continues up to 2015. And one can still find material on the internet saying the same thing: The eye's "poor" design shows that it is not the product of intelligent design but rather of blind evolution.[24]

But nearly twenty years before Dawkins's *Blind Watchmaker* (1986), the design advantages of the "backward" wiring had already been revealed. The rods and cones, it turns out, need plenty of blood to sustain their sensing end, and for this reason, if they were turned around to face the light, there would need to be a layer of blood between them and the incoming light. But since blood is largely opaque to light, it would block the vision. Wells explains this clearly in a short Discovery Institute video.[25]

There are many other marvelous features of the eye, too numerous to go through here. I will only add that the eye is sensitive enough to detect a single photon, although we need a stimulation of at least five to nine photons arriving within 100 milliseconds for the light to impinge on our consciousness.[26] Willful blindness is the only explanation for why the Darwinists hold on to their "flawed design" propaganda, which is a point that Wells makes in *Zombie Science*: Dead explanations keep getting resurrected to prop up a dead theory.

Not all "imperfections" have such a neat explanation as the eye. There is the laryngeal nerve in the giraffe, mentioned above, which goes from the brain down to the heart before going back up to the larynx. There is an ongoing debate about whether this detour makes any sense. The Darwinists' argument is based on the intuition that shorter is better. But is it, necessarily? There are some indications that this circuitous route can be defended in terms of design trade-offs, having to do with the constraints of embryological development and the role branches of the nerve play in several organs along its path.[27] I am not going to unpack the argument for and against in this case. I just take the opportunity to remind the reader that this convoluted route has made more and more sense as we have learned more and more about the laryngeal nerve, and that the Darwinists' reflexive assumption of bad design tends here, as it did with "junk" DNA, to function as a science-stopper. In contrast, a systems biology approach (inherently friendly to ID if not synonymous with it) assumes optimal or near-optimal design as a working heuristic; and it is proving extraordinarily fruitful and increasingly popular as a research framework.

Systems biologists, like engineers trying to reverse-engineer a piece of technology they don't fully understand, keep in mind that we often do not know as much as we think we know about what a particular biological part, system, or engineering choice is meant for. An organ or limb may be assumed to have only one function when it was designed to have two or three. Or perhaps the function of a part in one organism differs from its function in another. Not every wing is designed to fly. On a penguin, it turns out to be an excellent flipper. But what is the purpose of an ostrich wing? Balance? Display? A sign of its ancestors once being able to fly? We do not know. Such a situation is not unique to biology. Quite often archaeologists discover artifacts that they immediately recognize as designed although they cannot figure out what the item was meant to do. Perhaps it was just abstract art. But it was clearly the product of an intelligent mind.

Let us return to the eye to make another point about perfection. A more interesting objection to the present design of the eye might be to ask why God did not design chemistry differently so that the rods and cones would not need so much blood. That way, they could point into the light and the blind spot would be eliminated. Granted, the present "backward" design seems to be optimal given the limitations of chemistry, but if God is the master of all, could He not do a better job by designing a different universe?

First of all, it is easy to assert that a universe with a different chemistry could solve the problem, but even if we grant this unsupported assumption, what would the trade-offs be? As noted, the laws and constants of physics and chemistry appear to be exquisitely fine-tuned for life, which means that if we were to jigger one of these laws or constants to improve one situation, we likely would compromise a hundred other things.

But the ultimate response to such speculations is to recognize that if God were constrained to make the optimal universe, then He could never create. He alone is perfection itself, which by definition cannot be created since the being than which none greater can be conceived possesses being intrinsically rather than contingently, and anything God creates exists contingently. Also, any created universe would be finite and so could be improved upon in some way. Only God is infinite in that his being is in no way constrained. A universe might be eternal in time or in space, but it would still depend on God and be in some way finite.

That being said, Christian eschatology does make clear that the present creation will be surpassed when God makes "all things new," when He reveals "a new heaven and a new earth." Nevertheless, perhaps this present world represents a constrained optimization given His present purposes—purposes wiser, deeper, holier, and more mysterious than we can presently grasp. We cannot grasp them, but He has already left us some clues as to his reasons. The Garden of Eden was a marvelous home for man, and God wanted the best for humanity. And yet God drove them from the Garden into a world filled with pain, disease, and death. He did so not to be vindictive, but because, in his wisdom, he understood that for fallen humanity, the Garden was no longer good for us—nor was immortal life in our present fallen state.

Happily, that is not the whole of the story, nor is it the end. And what we can say for this vale of tears is that the infinitely wise God chose not only to create this world, but also to die on a cross to redeem it, meaning that at minimum it is a surpassingly wondrous possible world.

Summary

The cruelties and imperfections of nature do not invalidate the ID argument. But to make sense of their existence, we need to remember that we live in a fallen world and that somehow the whole of creation is touched by the rebellion of Adam and Eve, and perhaps also by the rebellion of Satan and his angels, a rebellion which precipitated a great cosmic battle between good and evil. Nor do the cruelties and imperfections of nature prevent us from seeking the God who can save us from its horrors and apparent futility. Fortunately, the cosmos is permeable to the action of God so that He can enter our world and lead us to Himself.

6. Prehistoric Man

Man is the result of a purposeless and natural process that did not have him in mind.
—George Gaylord Simpson, American Paleontologist[1]

At the inaugural Mass of his papacy, Benedict XVI said, "We are not some casual and meaningless product of evolution. Each of us is the result of a thought of God. Each of us is willed, each of us is loved, each of us is necessary."[2] It is hard to imagine that the newly minted Pope did not have Simpson's words in mind when he delivered his homily. If Darwinism were restricted to all life except human beings, it would only be an affront to common sense and empirical evidence. But it also offers a creation account of man, and one diametrically opposed to the book of Genesis and the Magisterial teaching about original sin. So Darwinism does not just offend common sense; it is a direct attack on Christianity.

We will look at the motivations of Darwin and his chief defenders in Chapter 8. Here the focus will be on what is currently known about prehistoric man.

In 2021, Sergio Almécija, the lead researcher of an extensive review paper on human origins, admitted, "When you look at the narrative for hominin origins, it's just a big mess—there's no consensus whatsoever."[3] Despite this caveat, it makes sense to give the reader some idea of what is being discussed in the field of prehistoric humans.

Before beginning, I want to clear up a bit of vocabulary. The literature lumps under "hominin" the following: modern man (*Homo sapiens*), various prehistoric humans (*Homo neanderthalensis*, *Homo erectus*, etc.), and primates thought to lead to modern humans or to be the close "cousins" of prehistoric humans (e.g., *australopithecines*). I will avoid using "hominin" and rather speak of both modern man and prehistoric man as members of the genus *Homo*.

The Fossils

Shortly after the appearance of *The Origin of Species* (1859), Thomas Henry Huxley published *Evidences as to Man's Place in Nature* (1863). The frontispiece featured an engraving of the skeletons, lined up from left to right, of a gibbon, orangutan, chimpanzee, gorilla, and man. It is said to be the inspiration for Rudolph Zallinger's 1965 illustration colloquially known as "The March of Progress," which in various revised forms has become one of the most celebrated icons of evolution. Huxley did not think that his sequence of skeletons was the actual line of descent from primitive primates to man, but he was arguing for an evolutionary origin for the human race.

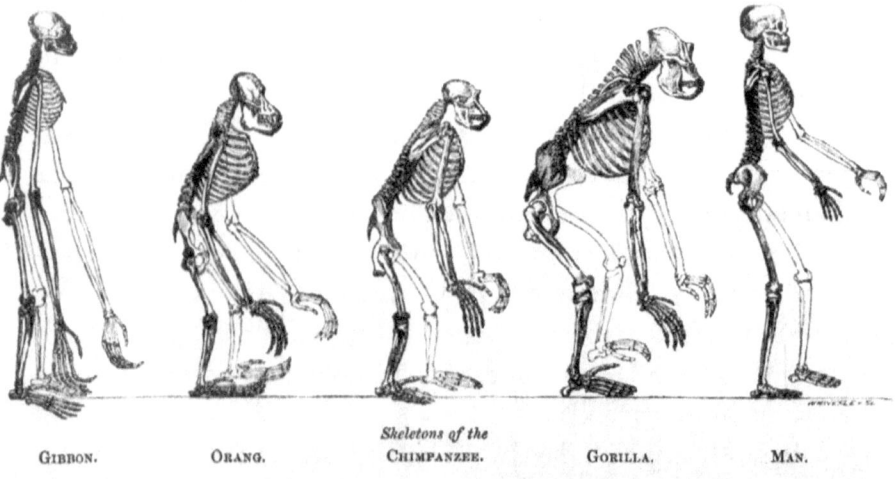

Figure 6.1.

Frontispiece to Huxley's *Evidence as to Man's Place in Nature* (1863). Note that the gibbon, a species much smaller than other four, is not drawn to scale with the others in the picture.

At the time of Huxley's writing, the only primitive human fossils, recognized as such, were of Neanderthal man, unearthed in 1856 in the Neander Valley, near Dusseldorf.[4] There was, at first, some argument as to whether the slight morphological differences between those bones and those of modern man were due to some pathology. Depending on one's stance in the debate, the bones belonged either to our species *Homo sapiens* or to a distinct species classified as *Homo neanderthalensis*. Later finds ruled out the pathology argument but, as we will see in this chapter, that is not to deny that Neanderthals were human. Altogether we now have fossils from over 300 individuals, from regions as far scattered as Israel, Iraq, Siberia, and Europe. The commonly accepted dating for these fossils is from 400,000 years ago to 40,000 years ago.[5]

In 1891, Eugène Dubois found a fossil that became known as Java Man. Dubois thought that it was a link between humans and apes and so gave it the name *Pithecanthropus erectus*, meaning "erect ape man." Later, the fossils were reclassified as *Homo erectus*, also sometimes called *Homo ergaster*, meaning "working man," because stone tools were found in the vicinity of some of the fossils. Later still, fossil bones from about forty individuals were found in China, which were known for a time as Peking Man, but, like Java Man, were later reclassified as *Homo erectus*. Since then, other fossils of this species have been found, the most famous being Turkana Boy, discovered in 1984, near Lake Turkana in Kenya. *Homo erectus* has been dubbed the first cosmopolitan lineage, because fossils have been found in Africa, the far east of China, Indonesia, Georgia, Italy, and Spain. The fossils are dated from 1.9 million years ago (1.9 Mya) to about 100,000 years ago (100 Kya).[6]

More recently, fragmented human remains were found in the Denisova Cave in southern Siberia. DNA analysis has determined that the fossils came from individuals who lived between about 50,000 and 285,000 years ago.[7] The jawbone of another Denisovan found in a Tibetan plateau has been dated to about 160 Kya.[8]

It may be that all of these are fully human in the theological sense of the word—beings created in the image of God, descendants of Adam and Eve. We will come back to that later. For now, we can be content to recognize a close skeletal affinity to modern *Homo sapiens*, which is something that the Darwinists do not dispute. The real difficulty for the Darwinians is to identify the ancestors of this branch of the evolutionary tree.

One obvious challenge with going back further in time is that there are fewer and fewer fossils of anything that could potentially be identified as a human ancestor. The problem is then compounded by the condition of the few fossils that are found.

Perhaps the most famous pre-*Homo* fossil, Lucy, was discovered in Ethiopia in 1974. It is dated to about 3.2 Mya and is classified as *Australopithecus afarensis*, meaning a "southern ape from the Afar region of Ethiopia." The name Lucy comes from the Beatles song "Lucy in the Sky with Diamonds," which was playing in the camp at the time of the finding. She became such a celebrity because approximately 40 percent of her bones were found, rather than just a bone or two (as is common for such ancient finds), although there is some dispute as to whether all the bones came from the same individual, and, in fact, one of the bones was proved in 2015 to have come from a baboon.[9] In any case, most of her cranium was missing and her pelvis badly damaged. Given this, it is not surprising that evolutionists disagree as to whether the *Australopithecines* were bipedal like modern humans, and how, when, and where exactly they evolved into our genus *Homo*.

Ardi, the fossilized remains of a female *Ardipithecus ramidus*, provides an even more dramatic example of some of the challenges paleoanthropologists face. The fossil, which dates from 4.4 Mya, was found in 1994 in Ethiopia and was widely trumpeted as the breakthrough discovery of a possible distant ancestor of humans. The name indicates the belief of her classifiers that her species was the root (*ramid* in Afar) near the ground (*ardi* in Afar) or foundation of apes (*pithēkos* in Greek, Latinized as *pithecus*) and, eventually, us.

Although the fossils of others of her kind have been discovered, Ardi is special because of the large number of bones found—125 in all, about half the total—and because these bones included parts of the pelvis, which made it possible to speculate about her bipedal abilities. Nevertheless, the bones were in such bad shape that it took fifteen years to assess them. A report in *Science* published in 2009

6. Prehistoric Man / **181**

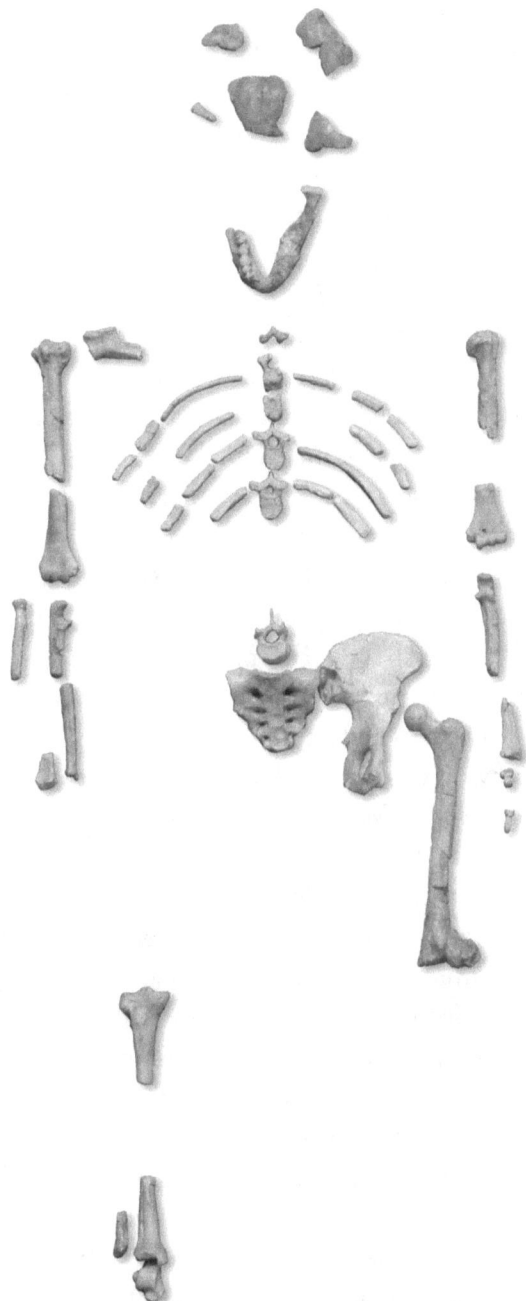

Figure 6.2.
Reconstruction of the partial fossil skeleton of 'Lucy,' *Australopithecus afarensis*.

acknowledged: "The team's excitement was tempered by the skeleton's terrible condition. The bones literally crumbled when touched. [Team leader Tim] White called it roadkill. And parts of the skeleton had been trampled and scattered into more than 100 fragments; the skull was crushed to 4 centimeters in height."[10]

Not surprisingly, there were some who doubted that Ardi was bipedal or a human ancestor. Esteban Sarmiento, a primatologist, wrote in *Science*, "All of the *Ar. ramidus* bipedal characters cited also serve the mechanical requisites of quadrupedality, and in the case of *Ar. ramidus* foot segment proportions, find their closest functional analog to those of gorillas, a terrestrial or semiterrestrial quadruped and not a facultative or habitual biped."[11] A study of the hands of Ardi came to the conclusion that they belonged to a creature more similar to chimpanzees and bonobos that spend most of their time in trees.[12] Stanford University anthropologist Richard Klein summed up the matter: "I frankly don't think Ardi was a hominid, or bipedal."[13]

A still earlier fossil, the Toumai skull, was found in Chad in 2001 and dated to about 7 Mya. In an article in *Nature* analyzing its significance, Bernard Wood wrote, "A single fossil can fundamentally change the way we reconstruct the tree of life."[14] He went on to explain that if *Sahelanthropus tchadensis* (the Toumai skull) is accepted as an ancestor of humans, then the whole tree as it stands will need to be reshuffled. In fact there are major aspects of the hominid tree that are totally unsupported by fossils. In 2015, Wood co-authored a major review of the hominid fossil record which admitted that "the evolutionary sequence for the majority of hominin lineages is unknown. Most hominin taxa, particularly early hominins, have no obvious ancestors, and in most cases ancestor-descendant sequences (fossil time series) cannot be reliably constructed."[15] This gives the lie to Ronald Wetherington's testimony to the Texas State Board of Education that human evolution has "arguably the most complete fossil succession of any mammal in the world. No gaps. No lack of transitional fossils... So when people talk about the lack of transitional fossils or gaps in the fossil record, it absolutely is not true. And it is not true specifically for our own species."[16]

Given the excitement about the Toumai skull, one might conclude that here was another relatively complete skeleton, such as Ardi or Lucy. But no, the initial study only reported a skull and some jaw fragments. Based on this and the presence of some animal fossils in the area, the paleontologists speculated about the lifestyle of *Sahelanthropus tchadensis*. In 2020, a study finally reported the femur of this species and predicted that it walked on all fours, like a chimp.[17] But given the overall paucity of clues, many of these speculations carry little weight. As the primatologist Frans de Waal points out, the skeletons of a chimpanzee and a bonobo are nearly identical, but the habits of the creatures are very different. The problems become greater when bones are missing. The anatomist C. E. Oxnard notes, "A series of associated foot bones from Olduvai [a locality bearing australopithecine fossils] has been reconstructed into a form closely resembling the human foot today although a similarly incomplete foot of a chimpanzee may also be reconstructed in such a manner."[18]

The reconstructions are often based more on the biases of the researchers than on anything truly objective. And the field is known for its strong personalities competing with one another to make a name for themselves and snag fresh grant money for future additional field research. In 2002, Mark Davis, who interviewed several paleontologists for a PBS documentary, remarked, "Each Neanderthal expert thought that the last one I talked to was an idiot, if not an actual Neanderthal."[19]

It should also be noted that when some new fossil is found, it is usually accompanied by an admission of previous ignorance which had been vehemently denied before. Geologist Casey Luskin, co-author of *Science and Human Origins*, cites the hype accompanying the find of a few teeth that were interpreted as linking *Ardipithecus* to *Australopithecus*. Apparently these few teeth from some 4 Mya made the "most complete chain of human evolution so far," which does not say much for the complete chain. And yet armed with this new "knowledge," the science reporter could write, "Until now, what scientists had were snapshots of human evolution scattered throughout the world."[20]

It might be useful to compare the anatomy of modern man to present apes to show that despite the obvious similarities, there are significant differences. We can begin with the location of the *foramen magnum*. The *foramen magnum* is the great hole in the bottom of the skull through which the spinal cord exits the cranium and meets the spine. In humans, the *foramen magnum* is in the middle of the skull, which enables the head to be balanced in an upright position. In apes, it is much further back.

Then there is the curvature of the spine and the change in the shape of the chest. The barrel chest of the primate gives it great strength for swinging through trees. Such a shape, however, could not be easily balanced by the upright walking human. The femur and the hip joint are also different in humans to make extended upright walking possible. The feet too need to be modified for walking upright. Human feet are relatively rigid and not prehensile, whereas chimpanzee feet are flexible and prehensile.

There are many other differences that one could cite. But these should suffice to show that the differences are significant and require coordinated changes.

There are several hundred fossils of prehistoric humans. But there are no clear connections or links which prove the connection of these humans to earlier species of non-humans such as the australopithecines. Luskin cites many mainstream scientific sources acknowledging the lack of fossils documenting any sort of evolutionary transition between the apelike australopithecines and the humanlike members of the genus *Homo*. Many examples could be given, but one of the more poignant quotations comes from Ernst Mayr, who noted that the "earliest fossils of *Homo*... are separated from *Australopithecus* by a large, unbridged gap."[21]

Moreover, everything that has been said in a previous chapter about the impossible odds that natural selection has to face applies in the case of human origins. The changes in skeletal structure needed for upright walking—the femur with the angled femoral head, the repositioning of the backbone towards the center of the head, the enlarged knee, and more[22]—are all complex changes, many of which need to occur in tandem with other changes to confer any advantage. This poses a severe challenge to gradual Darwinian evolution, a challenge compounded by the fossil record not recording any such series of small, step-by-tiny-step changes.

DNA and Human Origins

Molecular biology is a relatively recent tool to probe prehistoric man. It was not until the 1950s that DNA was clearly identified as the molecule that contains much of the information necessary for life. Genome sequencing started slowly in the 1970s and came into its own only after the turn of the millennium. Each year, the techniques get more powerful and less expensive. It is now possible to examine DNA found in the bones of Neanderthals and other early humans. The present record for the oldest human DNA is over 400,000 years old from *Homo heidelbergensis* found in Spain.[23] DNA decay will not allow the record to go much beyond that. And one must remember that truly fossilized remains, i.e., those in which the organic matter has been replaced by minerals, will not yield any DNA data. There are several perspectives on the human story that molecular biology can provide, some of them sensationalized by the media. So let us take a look at these in turn.

i) DNA—Comparison to Present Primates

One often hears that we are 99 percent similar in our genes to chimpanzees and bonobos.[24] Such statements are usually made by those who think that we are nothing but a slightly different species of ape. In fact, that figure has begun to drift downward as genome sequencing and comparison techniques improve. A more up-to-date assessment would place the percentage of commonality somewhere in the mid-80s to mid-90s.[25] But even that much overlap is surely curious. The obvious rejoinder to such statements is that if indeed our DNA is so very similar to that of chimps, then it is clear that the deep and fundamental differences between us and chimps must lie somewhere outside of gene sequences. Part of the differences may rest in the ways the genes are expressed, and in the epigenome—the domain of a fast-developing field of molecular biology called epigenetics. But there is another source of the difference, and surely the most significant: the immaterial human soul.

One should not be surprised that our genetic make-up should resemble that of today's apes. Anyone with a pair of eyes can see that there are many physical similarities between us and the apes. So if the genetics have a lot to do with the physical makeup of the organism, the DNA is bound to be similar as well.

An interesting fact that is often brought up is that humans have 23 chromosome pairs whereas other primates have 24. The explanation adduced for this is that two chromosomes that are separate in other primates fused to become chromosome number 2 in humans. The fusion, it is said, happened after the branches of the Darwinian tree leading to humans and other primates split. So it is seen as further evidence of evolution. This, however, is convincing only to those who are eager to see common descent as the only explanation, because the location of the supposed fusion does not have the typical telomere sequences, sequences found at the ends of chromosomes,[26] but rather sequences that are highly degenerate, much more so than what fused telomeres would be expected to look like.[27]

Darwinians also argue that humans and other primates possess the same "pseudogenes," a pattern well-explained by evolutionary common descent but odd and surprising if humans are the work of an intelligent designer. After all, these pseudogenes are damaged genes, the argument goes, something we supposedly can safely conclude because they look similar to other genes that code for proteins but do not themselves code for proteins. Thus, it would seem that there are broken genes shared among primates. Why would a wise designer repeatedly stick a broken gene in one primate creation after another?, the thinking goes. He wouldn't. But blind evolution well might create the pseudogene via random mutation and then pass it along up the branching tree of life to various primates splitting off from the original ancestor, stuck with the freshly minted pseudogene. The problem with this tidy story is that recent research is revealing widespread function in pseudogenes.[28]

ii) DNA—Mitochondrial Eve and Y-Chromosome Adam

The typical cell in a human being has two types of DNA. There is the nuclear DNA, which is arranged in twenty-three pairs of chromosomes. The sperm contributes twenty-three chromosomes; and the egg contributes the other twenty-three. In addition, each cell has mitochondrial DNA, which comes from the egg alone. This mitochondrial DNA is produced in each cell and eventually in a new egg by straight copying. There is no "meiosis dance" to mix up the genetic material as happens in the case of the production of the twenty-three chromosomes that will be in the sperm and egg.

The small circular chromosome of mitochondrial DNA is much shorter than the nuclear DNA, some 16,500 base pairs as compared to three billion. So the first genetic analysis of human history focused on mitochondrial DNA. In 1987, Allan Wilson and his colleagues sequenced stretches of this DNA obtained from people scattered throughout the world. They found that there was more variation among individuals in Africa than among individuals in other parts of the world. This makes sense, if one believes that humans originated in Africa and later migrated to other parts of the world. Based on an average mutation rate—a molecular clock—Wilson surmised that the most recent common ancestor (MRCA) of all who are alive today lived in Africa about 200,000 years ago.[29] More recent estimates date the mt-MRCA (the "mt" stands for mitochondrial) to between 156,000 and 120,000 years ago.[30]

Once the cost of genome sequencing went down, it became possible to analyze the Y-chromosome in males throughout the world to date the earliest Y-MRCA. (The Y-chromosome, like the mitochondrial DNA, is also passed on without meiosis mixing things up.) The estimated date for this male from whom we are all descended was 99–148 Kya, according to one study, and 180–200 Kya ago according to another. And once again, in Africa.[31]

In popular parlance mt-MRCA and Y-MRCA are known as Mitochondrial Eve and Y-Chromosome Adam. Despite the biblical names, there is no claim that these two were the first human beings. Nor is it claimed that they actually formed an actual couple who lived at the same time. These are merely the most recent to whom we can all trace our lineage via two parts of our genome—the mitochondrial DNA and the Y-chromosome.

Should this at first seem confusing, see the figure below, which illustrates how other progenitors can get totally wiped out from the gene pool. It may be helpful to illustrate this with a simplified scenario. Think of the biblical Adam and Eve being created one or two million years ago, then a great catastrophe happens two hundred thousand years ago and wipes out all humans except for one couple on a remote island. These then go on to reproduce and repopulate the earth. A DNA study based on the samples of today's humanity would pinpoint this Y-Chromosome Adam to have lived two hundred thousand years ago rather than the million or two years ago when the first human couple was created. Next, a hundred thousand years ago, a strange disease wipes out all the women in the world except for one. All the human beings in the world now living could be traced back to this Mitochondrial Eve, who lived 100,000 years after Y-Chromosome Adam. But note that even in these two extreme scenarios, the DNA of humans from as far back as Adam and Eve one or two million years ago contributed to the genome of these widely spaced individuals.

Of course, the actual process of ending up with most recent common human ancestors who lived so much more recently than the first human couple undoubtedly was vastly more complex than my simplified scenario. But the scenario should help us to see that these Mitochondrial Eve and Y-Chromosome Adam findings do not rule out that there are other sections of our DNA that come from ancestors still further back in time or that Mitochondrial Eve and Y-Chromosome Adam need not have been contemporaries.

iii) DNA—The Possibility of Monogenism

The Catholic faith obliges us to believe that all human beings are in fact descended from Adam and Eve. What does science have to say about this proposition? If one is to believe the late Francisco Ayala, an apostate Catholic priest and biologist, the present variations in the human genome indicate that at no time was there a population of fewer than 10,000 individuals which gave rise to what we now call humans. Biologist Ann Gauger has analyzed his mathematical study and found it wanting in its assumptions. In fact, by the time she

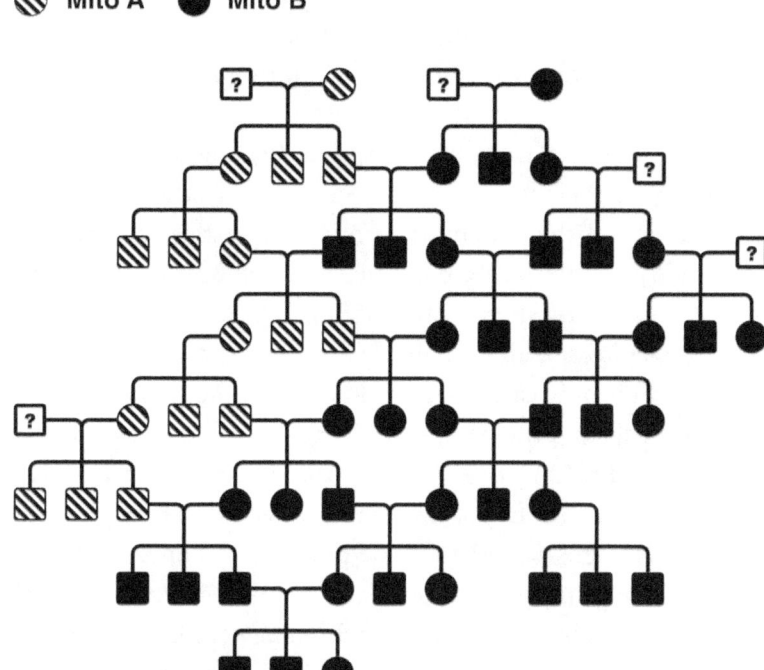

Figure 6.3.

Circles represent females and squares, males. This simplified genealogical table illustrates how Mitochondrial "Eve" (Mito B, the uppermost solid circle) need not be the Eve of Genesis, the first woman. The mitochondrial lineage of the Mito A woman (uppermost striped circle) ends when the one remaining female in that line has only male children (the row of three striped squares). Beginning in the seventh generation of the A and B lineages, all the descendants, male and female, trace their female line exclusively back to Mito B. Geneticists can trace female lineages because mitochondrial DNA is passed exclusively from mothers to their children, with no input from the fathers. (Question-mark squares signify men whose mitochondrial lineage is unknown.)

finished her analysis, she was convinced that two individuals could account for all the variation in a particular kind of gene on chromosome 6, the one that Ayala had chosen as his subject.[32]

Meanwhile, the genetic diversity argument touted by Ayala and others against Adam and Eve was repeated by Dennis Venema,[33] a biologist affiliated with BioLogos. (BioLogos aims to convince

Christians to accept Darwinism.) In 2017, Richard Buggs, an evolutionary geneticist at Queen Mary University in London, pointed out to Venema and other scientists on the BioLogos discussion forum that there was a problem with their argument. Venema graciously admitted his error. In the meantime, Ann Gauger and Ola Hössjer, a professor of mathematics at the University of Stockholm, conducted further studies on the possibility of human genetic diversity arising from a single pair, which they published between 2016 and 2019. Their conclusion was that a single couple living 500 Kya or more could be the sole progenitors of all humans. Joshua Swamidass, a theistic evolutionist, did his own calculations and came to the same conclusion.

It is a credit to Venema and BioLogos that they admitted their error. As Casey Luskin writes, "BioLogos's president, Deborah Haarsma, acknowledged BioLogos's past mistakes on this topic, including that some BioLogos scientists 'made premature claims … that evolutionary science and population genetics rule out scenarios with a recent universal human ancestor or with a *de novo* created ancestral pair.' She further wrote: 'Over the years, we have removed old content from our website for many reasons, including articles… that overstated scientific claims, that unnecessarily excluded theological positions that are consistent with scientific evidence… we need to be honest when we overstate an argument.'"[34]

iv) DNA—Ancient Human Genomes

DNA sequencing has come a long way since the early studies on Mitochondrial Eve. It is now possible to sequence the DNA of early modern humans and even Neanderthals and earlier humans. David Reich, one of the pioneers of this field, has written a very readable account of the state of the research. He prefaces his work of popularization with a caveat that the research is moving so quickly that some of the things that he writes might need a major revision by the time that his book is in the press. Nevertheless, he feels confident enough to offer some general observations.

First, the new discipline reveals the need to correct the simple "out of Africa" scenario. By this account, developed in the 1980s and 1990s, modern man originated in Africa about 150,000 years ago and went on to colonize the world. The mitochondrial DNA evidence fits quite nicely with this version. And Reich does not claim that the "out of Africa" scenario is completely wrong, but rather that it needs to be modified.

He argues that anatomically modern humans (*Homo sapiens*) may have existed for up to 300,000 years and that, following the genetic evidence, we also have among our ancestors Neanderthals and Denisovans, who interbred with our modern human forebears.

According to Reich, the lineage leading to the Denisovans split from the lineage leading to Neanderthals and modern humans at some point between 1,000 and 800 Kya; and the lineages leading to modern humans and Neanderthals split at some point between 770 and 550 Kya. Modern humans then mated with Neanderthals at some point between 54 and 49 Kya; and with Denisovans between 49 to 44 Kya.[35] These unions left genetic traces. Reich presents a map showing how much Neanderthal DNA and Denisovan DNA people in different parts of the world now have. In sub-Saharan Africa the percentage is said to be zero, but in some sections of Europe and Asia, the Neanderthal percentage is estimated at around 2 percent. There is said to be a similar amount of Denisovan DNA in southeast Asia, in particular on the island of New Guinea.

So, according to Reich, the modification to the "out of Africa" scenario is that there were several expansions out of Africa. The oldest migration gave rise to *Homo erectus* going into Asia and Europe. Through isolation of groups, this early species of man diversified, and developed the specific genomes associated with the Denisovans and Neanderthals. These then interbred with one another. And they interbred with another out-of-Africa migration that took place less than 100,000 years ago.

Reich believes that the new research already gives much information about the migration and mingling of human populations in the last several hundred thousand years. But he has doubts that it will be able to shed much light on human biology.[36]

This prognosis is a disappointment to earlier hopes that a genetic switch could be found to explain the flowering of human cultural artifacts beginning about 50 Kya. Darwinists were hoping that some major biological changes could be accounted for by relatively minor genetic changes. For example, the change from wild teosinte into maize required a mere five genetic switches. By analogy, early researchers thought that perhaps a minor modification of a gene associated with the ability to speak, the *FOXP2*, could account for the great cultural leap forward associated with *Homo sapiens*. But one problem with this hypothesis is that the new research has shown that modern humans and Neanderthals shared the *FOXP2* mutation, which is not found in chimpanzees. (More accurately, this is a problem for those who deny that Neanderthals had comparable intellectual abilities to *Homo sapiens*, something that will be discussed in the next chapter.)

There is also an in-principle problem with the idea. The transition from teosinte to maize is one thing. Going from the mental capacities of ape-like creatures to human beings is an exponentially greater leap—a qualitative difference without parallel in the natural world. If DNA alone were responsible for the difference, we have every reason to assume that far more than a modification to a single gene is necessary, since in every other realm of biology (as well as in human engineering) sophisticated novel functions require enormous amounts of novel information. In fact, the complexity of human language and the unlikelihood that a single gene could be responsible for it was recognized by the journal *Nature*:

> Language is complicated, and was never going to be explained by a single mutation in modern humans, [Simon] Fisher adds. "We need to embrace more-complex accounts that involve changes of multiple genes. In that sense, *FOXP2* was only ever going to be one piece of a complex puzzle."[37]

Clearly the origin of human language cannot be explained by small changes in a single gene. There is already a lot more to look at. Apparently genome sequencing of Neanderthal DNA has now identified more than 100,000 places that differ from the common human genome of today. And as I will argue later, the real leap from non-human to human may have occurred much earlier, as far back as two million years ago, with the origin of *Homo erectus*. If so, the thousands of genetic differences between the genomes of *Homo sapiens* and Neanderthals may all be cases of microevolution.

Reich surmises that "it will take an evolutionary Manhattan Project to understand the function of each mutation we have and that Neanderthals do not." Further, he speculates that if the project is done, "the findings will be so complicated—with so many genetic changes contributing to what makes humans distinctive—that few people will find the answer comprehensible. While the scientific question is profoundly important, I expect that no intellectually elegant and emotionally satisfying molecular explanation for behavioral modernity will ever be found."[38]

Christians, of course, should not find this surprising. What makes humans unique is the spiritual soul. No doubt the human body was specialized in its genetic makeup to enable it to form a functional unity with the soul. But the molecules themselves are not going to come close to providing the most important answers as to what makes humans created in the image of God.

Evidence of Culture

The big puzzle in early human history is the lack of cultural artifacts that go back much further than 50,000 years. One can, on the basis of anatomy, the genome, and even brain sizes argue that *Homo erectus* (and many other members of the genus *Homo*, such as *Homo heidelbergensis)*, the Denisovans, the Neanderthals, and *Homo sapiens* were all human beings in the theological sense—that is, made in the image of God and descended from Adam and Eve. It is perhaps a bit humbling that the average Neanderthal's brain, at 1450 cc, was slightly larger than ours, at 1345 cc. The average brain size of *Homo erectus* was smaller at 1016 cc, but given that the brain sizes of modern humans vary between 800 and 2200 cc, that is clearly still big enough to be human. Chimpanzee brains, as a comparison, vary from 275 to 500 cc.[39]

If the surmise is true that *Homo erectus* was the first human being, one might expect to see evidence of culture beginning over a million years ago. But in the first million years that *Homo erectus* roamed the earth, we do not see anything beyond an advance in stone toolmaking and a handful of other, usually ambiguous, signs of primitive technological progress, such as the possible use of fire, and even here it would seem to be not the capacity to reliably start and maintain fires but more likely just the opportunistic use and limited control of fire in the wake of naturally occurring fires. Why so little progress—or more precisely, why so little archaeological evidence of progress—for more than a million years in the fossil record if *Homo erectus* were fellow humans, made in the image of God?

As noted, the archaeological record is, as one might expect, spottier the further one goes back, but there are a host of intriguing clues pointing to slow technological progress beginning with *Homo erectus* and eventually accelerating, first gradually and then explosively. Let's take a few pages now to dive deeper into these clues.

i) Use of Tools

The use of tools is not limited to humans. Birds use straws to get at bugs; sea otters use rocks to crack open abalone shells; chimpanzees use stones as hammers; and there are yet more examples of animals using simple tools. But humans took tool-making to new heights.

The archaeological record has stone tools dating from at least 3.3 Mya.[40] Some of these examples are of dubious status as tools, but they may be primitive hammerstones or sharp cutting instruments. These predate humans, but this is not too surprising; when a modern chimpanzee uses rocks to break open nuts, he is similarly using a primitive hammerstone. Later we find the Oldowan toolkit, from the Olduvai Gorge, where they were first discovered.[41] By 1.7 Mya (and possibly as early as 1.95 Mya[42]) there are larger stones worked so as to produce a large cutting edge, which can be called a hand-axe. These are the basics of what is known as the Acheulean toolkit, named after the town of Saint Acheul (France), where they were first discovered in 1847. Since then, they have been found in Africa, Europe, and Asia. They continued to be widely used until as late as about 130 Kya in some places. The period saw some refinements in technique, but the basic strategy remained consistent, allowing archaeologists to group this long period of the Stone Age under the term Acheulean or Mode 2.[43]

A period stretching from the appearance of the most primitive stone tools (predating the earliest *Homo erectus* fossils) until the development of metal tools is broadly referred to as the Stone Age. This leaves the impression that stone was the only tool material used by our ancestors from this ancient period. But the term is meant only to indicate that this period predates the invention of metal tools. There is also archaeological evidence of wood[44] and bone[45] tools stretching back hundreds of thousands of years into the Stone Age. And since wood and bones small enough to function as hand tools cannot be expected to preserve as well as stone over the course of many hundreds of thousands of years, it's reasonable to suppose that the absence of still older wood and bone tools may be a preservation problem rather than a function of their not having been invented yet.

About 200 Kya, there is an acceleration in the design of stone tools. One can find sharp stone points that can be attached to sticks so as to make spears or arrows. And there are awls to perforate animal hides and scrapers that could be used for a variety of purposes. These are examples of Middle Stone Age tools and are found until about 50 Kya.[46]

The Later Stone Age tools include needles for sewing and harpoons with barbed points made out of bone, as well as tools made from wood and stone such as axes with wooden handles.

ii) Use of Fire

As noted above, there is a difference between sustaining fire kindled by lightning and starting a fire. The earliest findings of evidence for fire use by *Homo* tend to be burnt bones and some layers of burnt material, which makes it very difficult to evaluate whether the fire was even used by humans, let alone started by them. There are several sites dating to between 1.5 Mya and 700 Kya, including two cave sites in southern Africa and Gesher Benot Ya'aqov in Israel, along the Jordan River to the north of the Sea of Galilee.[47] The latter site includes tiny burnt flints and burnt wood. Use of fire may stretch back even as far as 1.5 Mya, if evidence at Swarthkrans cave in South Africa holds up.[48]

The Zhoukoudian cave complex site near Beijing was first discovered in 1921, yielding Peking Man. For about eighty years now there has been debate as to whether the caves provide evidence for the use of fire. There is some burnt bone, but it is not clear whether the fire was from uncontrolled natural sources or from constructive human use. There is also debate as to the time period in which the *Homo* habitations date from, with the range given as about 770 to 400 Kya.[49]

There is more evidence of the use of hearths dating from 400 Kya and more recently in just about all parts of the world colonized by ancient humans that early.[50] But researchers in the field acknowledge the relative scarcity of these finds. And they note that some sites containing much in the way of bones and tools show no signs of fire.

Pierced shell beads, dating from about 120 Kya, found in Israel and Algeria, indicate that twine or leather cord must have been known, which in turn indicates that the elements for starting fire by friction were in place. After these dates, fire is almost universal. Further impetus to the use of fire came about through the discovery of pottery, some 20 Kya in China. And beginning with the development of agriculture, some 12 Kya, fire was used to clear land.[51]

iii) Metal-Working and Glass

Metallurgy, with the exception of native copper, could take off only after early man learned to control fire. At first metals like copper were used in their native state, that is unsmelted, perhaps as early as 8000 BC in the Middle East.[52] Copper smelting dates from around 5000 BC in the region that is today's Serbia.[53] But of more significance is copper-tin bronze, which dates from about 3200 BC in Asia Minor, heralding the Bronze Age, which began about 3000 BC.[54] There is evidence of iron smelting as early as 2500 BC in today's Turkey.[55]

The earliest manufactured glass, as distinct from naturally occurring obsidian, goes back to at least 3600 BC,[56] which is near the time that bronze appeared. And, in fact, it has been proposed that early glass may have been a serendipitous product of metal production with its higher-temperature fires.[57]

iv) Art: Wall Paintings and Statuettes

Although chimpanzees, otters, and many other animals might use primitive tools, only humans produce art. The first prehistoric paintings to be discovered in modern times were in the Altamira Cave in the Cantabrian Mountains in 1879. They were at first thought to be forgeries because of the skill of the artists and lack of soot on the walls; but by the beginning of the twentieth century, they were accepted as prehistoric. They are now thought to have been created over many years. The oldest picture is thought to be from about 36 Kya, and the most recent from about 22 Kya.[58]

Figure 6.4.
Replica of lion painting in the Chauvet Cave (Ardèche, France).

Many people are familiar with the wall paintings in the cave at Lascaux, in the Dordogne region of France. Discovered in 1940, the cave contains nearly 600 color paintings of various animals. The images date from around 17 Kya. The Chauvet Cave, in the Ardeche region of France, was discovered more recently, in 1994. It contains pictures and engravings of animals, which are just as stunning as the pictures at Lascaux, except that the Chauvet artists produced their work much earlier, about 36 Kya.[59]

Recently, three caves scattered across Spain were found to contain paintings that were dated to more than 65 Kya.[60] And in one of these caves, in southeastern Spain, researchers found perforated seashells and pigments dating to at least 115 Kya.[61] The surmise is that these objects were produced by Neanderthals, because modern humans did not arrive in Europe until about 45 to 40 Kya.[62]

Neanderthals also are credited with the constructions of circular rings made of stalagmites. Near the village of Bruniquel in southern France, there is a cave that contains these structures, along with some burnt bones and other evidence of heating, which dates to some 176 Kya.[63] The fact that these structures were over 300 meters from the entrance indicates that the builders had to have mastered a means to light the environment. As Jacques Jaubert and his colleagues report in *Nature*:

> The attribution of the Bruniquel constructions to early Neanderthals is unprecedented in two ways. First, it reveals the appropriation of a deep karst space (including lighting) by a pre-modern human species. Second, it concerns elaborate constructions that have never been reported before, made with hundreds of partially calibrated, broken stalagmites (speleofacts) that appear to have been deliberately moved and placed in their current locations, along with the presence of several intentionally heated zones. Our results therefore suggest that the Neanderthal group responsible for these constructions had a level of social organization that was more complex than previously thought for this hominid species.[64]

As Rebecca Wragg Sykes remarks in *Kindred: Neanderthal Life, Love, Death and Art*, "Bruniquel laughs in the face of austere, survival-only explanations for Neanderthal behaviour. It surely was made by thinking, but also *feeling*, minds."[65]

On the other side of the world, on the island of Sulawesi in Indonesia, there exists a cave that has stencils of hands, dating to about 40 Kya.[66] And there is a fragment of a cave painting in Australia, dating to 28 Kya.[67]

Besides cave paintings, there are petroglyphs in the La Ferrassie cave in France thought to have been inhabited by Neanderthals from about 70 to 50 Kya.⁶⁸ And in South Africa, at the Blombos Cave, there is a stone with geometrical etching thought to be from 73 Kya.⁶⁹

There are also many examples of small figurines of naked women referred to as Venus. There is the Venus of Hohle Fels or the Venus of Schelkingen, found in Southwestern Germany. It is made of ivory and is thought to date to about 40 Kya.⁷⁰ A much earlier statuette in stone, the Venus of Tan-Tan, was found in Morocco and reportedly dates from between 500 and 200 Kya. Admittedly that is a great spread of dates, and there remains controversy around the question of how much of the Venus of Tan-Tan's shape is due to natural weathering versus the intentional activity of an agent,⁷¹ so one has to be careful as to what to think of it. Besides these Venus statuettes, there is the Lion-man of Hohlenstein-Stadel in Swabia. It is an ivory carving about twelve inches high, thought to date from 40 to 35 Kya.⁷²

v) Musical Instruments

The earliest artifact that is thought by some to be a musical instrument is a bear bone with two linearly aligned holes. It was found in northwestern Slovenia and is thought to be 43,000 years old. Some think that it was made by Neanderthals. Others think that the holes were due to animal bites. In any case, a modern replica of the instrument was made and a professional musician was able to produce some very recognizable music.⁷³

Less controversial are two flutes from a cave in southern Germany. One is made from the bone of a bird and the other is from mammoth ivory. They are dated to between 43 and 42 Kya.⁷⁴

vi) Clothing

Archaeologists at first guessed that the earliest clothing was made from animal skins around 100 Kya. This was based on finding rock scrapers of that age, which they believed were used to scrape flesh from skins so as to make the skins suitable for wearing.⁷⁵ In 2011, the date for the earliest clothing was revised to potentially as early as about 170 Kya by studying the differences in the genomes of head lice and clothing lice. The date is the best estimate of when the two lines of lice diverged.⁷⁶

vii) Burying the Dead

Some early human bone pits date to over 300 Kya. The earliest undisputed intentional burials date to 130 Kya. There is a grave of Neanderthals in northern Croatia that dates to 130 Kya, although the burial hypothesis is controversial.[77] A Neanderthal "flower burial" gravesite was found in a cave in Iraqi Kurdistan dating to 70 Kya.[78] The first undisputed evidence of the burial of anatomically modern humans dates from about 92 Kya in Qafzeh in Israel. The bones were stained with red ochre, a mineral used in subsequent millennia in Africa.[79] The same mineral was found in what is believed to be a ritual burial of a man in Lake Mungo, Australia, some 42 Kya.[80] The oldest known human burial in Africa dates to about 78 Kya.[81]

viii) Prehistoric Jewelry

The earliest extant jewelry dates from about 100 to 135 Kya from Skhul in Israel, in the form of pierced seashells, perhaps strung together on a string that has long since disappeared.[82] Perforated shells dating from 115 Kya have been found in a cave on the southeast coast of Spain, and are attributed to Neanderthals.[83] There are ostrich-egg beads from Tanzania dating to about 52 Kya.[84] And at the Sungir site, about 200 kilometers east of Moscow, there are three significant burials dating to about 34 Kya, which include bones filled with red ochre and skeletons adorned with over 13,000 mammoth ivory beads. These would have taken an estimated 10,000 or more hours of labor to produce.[85]

ix) Language

Language is one of the most characteristic attributes of humanity. Historians think writing might have been developed independently in several different parts of the world, with all these developments dating from after about 4000 BC—well after humans had colonized much of the world. Spoken language, in contrast, is generally believed to have emerged far earlier.

In the 1960s, Luigi Luca Cavalli-Sforza started work on the genetics of the human race. He took blood samples from people in nearly every part of the world and tried to estimate their hereditary distance from one another. The linguists Joseph Greenberg and later Merritt Ruhlen tried to classify languages for which written texts exist into large families. When they did this, there was a great similarity between their classification and the genetic classifications of Cavalli-Sforza.[86] David Reich, though noting that some of Cavalli-Sforza's results have to be modified in light of more recent whole-genome studies, nevertheless comes to the defense of Greenberg and his classification of native American languages into three groups. The classification aligns very well with genetic results that indicate that there were three major immigrations from Asia into the Americas.[87] If there is any merit to these studies, they suggest that humans were using language before they entered the Americas over 10 Kya.

Although the work of Greenberg and Ruhlen is disputed by many linguists, most archaeologists believe that man was able to speak by the time that he produced symbolic art. They present scenarios as to how language might have been developed around campfires and in activities needing collaboration. Or they pin their hopes on some genetic mutation in the *FOXP2* gene, but the problems with this were already mentioned earlier in the chapter. We will return to the problem of language in Chapter 7.

Summing Up—Prehistoric Man According to Modern Science

There is no convincing, coherent, evolutionary story of the development of the human race. The idea of a gradual, step-by-step Darwinian transition from ape-like predecessors to *Homo erectus* is based on the scantest fossil evidence and a lot of wishful thinking and highly imaginative, sometimes patently misleading misconstructions on the part of Darwinists. The timing of the arrival of the first true humans is also based on evidence that is often disputed, such as the average rate of change of the genome or the dating of soils and rock around fossils. The latter presents great problems if there are no radioactive elements nearby to help. Carbon dating cannot go beyond about 60 Kya and cannot be used on fossilized bones. Just about every element in the brief survey of the cultural artifacts of prehistoric man has been disputed as to its date or significance. All the researchers who do not believe in man's possessing a spiritual soul, which is to say the vast majority, are puzzled as to what caused the indisputable leap forward in human cognition and technical ability. But they keep trying to devise some plausible way to explain how a primitive animal groped its way to fully human consciousness.

That is not to say that the Church has a tidy answer as to why it seems that the first appearance of the human body, as evidenced by fossils, should predate by many millennia the evidence for an intellectual flowering. Are the dates wrong? Is the evidence for culture missing? And are there any solutions to this puzzle in what Christian faith suggests to us about prehistoric man?

Adam and Eve—Epistemic Tension

If we are willing to accept (a) that the soul of each human being is a spiritual reality, directly created by God, and (b) that there is no credible missing link between other primates and man, we are still faced with some challenges in understanding the early history of the human race. As mentioned, the main problem is that although the fossil evidence of the genus *Homo* goes back to some 2 Mya, the cultural flowering of man as evidenced in the Chauvet cave paintings and some others date from some 50 Kya.

To be precise, the picture is somewhat more complex than that. There are some paintings and statuettes that take us back further, to some 100 Kya, and the Bruniquel cave arrangements take us to 176 Kya. Additionally, there is some evidence of fire use at 1 Mya, and perhaps still earlier. Also, the earliest Acheulean (Mode 2) stone tools date to around 1.95 Mya, roughly coincident with the emergence of *Homo erectus*. While not dramatically different from the preceding Oldowan (Mode 1) stone tools to the untrained eye, they do mark a distinct technological advance over Mode 1 stone tools, and an advance that lies well beyond the capacity of chimpanzees.

But even with all this, one has to wonder, if the members of *Homo erectus* were human, made in the image of God, why were they stuck on this single stone-tool advancement for a million years or more? Perhaps they also developed a limited capacity for controlling fires at some point in their first million years of existence, but not much else. Why does the great cultural flowering come so much later?

Another potential problem for the biblical account is the large number of different species in the genus *Homo*. *Homo sapiens* is currently the only extant species, but archaeologists recognize *Homo antecessor*, *Homo erectus*, *Homo ergaster*, *Homo floresiensis*, *Homo habilis*, *Homo heidelbergensis*, *Homo luzonensis*, *Homo rudolfensis*, *Homo naledi*, and *Homo neanderthalensis*. In addition, they recognize subspecies such as the African *Homo erectus*, the Asian *Homo erectus*, and the Denisovans. Are all these, or at least some of these, descendants of Adam and Eve?[88]

Further, what are we to think of the biblical accounts? Monogenism implies incest, which certainly makes us feel uneasy if not outright disgusted. Yet, Pius XII insisted that we must hold on to a single original couple as a revealed truth, for there does not seem to be another way to explain original sin. If anything, reading the Bible more carefully just raises more questions. For example, in Chapter 4 of Genesis, we seem to be dealing with a populated world in which Cain fears for his life after slaying Abel. We are then presented with some genealogies that give hints of various stages of human cultural development—living in tents, raising cattle, playing the lyre and pipe, forging bronze and iron. How did we get to the Iron Age so quickly in the narrative?

Nor do the mysterious elements end there. In Chapter 5, we have another brief account of the creation of Adam, in which it is mentioned that Adam had other sons and daughters after the birth of Seth, who himself was born after Cain and Abel. All well and good. But then, in Chapter 6, we encounter "the sons of God," who took to wife "the daughters of men," with the offspring becoming known as the Nephilim, which in Hebrew can mean "giants" or "fallen ones." This somehow aroused God's displeasure, for he declares in verse 3, "My spirit shall not abide in man for ever, for he is flesh, but his days shall be a hundred and twenty years." A common way to read this has been to take God as saying that the long lifespan for man, which up until that time reached as high as 969 years, would be shortened to 120. But then in Chapter 7, we are told that the rains which caused the Flood came in the 600th year of Noah's life. And post-Flood genealogies have various individuals living well past 120.

OK, but who exactly were the Nephilim? And why are they said to be around after the Flood, if the Flood was global and wiped out everyone but the eight on the ark? And what's behind the great ages of men listed in much of Genesis? While biblical scholars have suggested a variety of possible answers to these questions—some more credible than others—the Church has not provided us with authoritative interpretations or even ruled on whether the Flood was local or universal.

It would take us too far afield to try to make sense out of all this in these pages. And the result would surely seem forced. It is mind-boggling that Bishop Ussher should have felt so confident in understanding the sacred text as to assign an exact day in 4004 BC as the creation of the world based on biblical genealogies. But Catholics do have to believe in monogenism and consequently that incest was necessary for the first few generations. If it helps, we might remind ourselves that perhaps sin had not immediately struck such a deep root as to necessitate the prohibition of incest for protection of the individual family members. Perhaps the genome would not have been as degraded as ours is now, so that the begetting of sickly offspring was not the danger it is today.

Let us return to the scientific evidence and, in particular, the problem of the various species and subspecies of the genus *Homo*. The Polish Jesuit scholar Piotr Lenartowicz notes that in the last ten thousand years we can distinguish between four or five natural races or ecotypes of man. "The ecotypes manifest an inner, developmental and adaptive potential of mankind," he writes. "One might ask if it is possible that during the glacial epoch or earlier, in the Pliocene epoch, the adaptive potential had worked in the same way but the range of its manifestations was proportionally broader."[89]

Classification is always a tricky business. Today we rightly classify all people as *Homo sapiens*, and regard the sometimes dramatic differences in height, weight, skin color, and facial features as variations within our wide-ranging species. Archaeologists, looking at the bones of various *Homo* groups, are freer to break them up into what they regard as separate species, even maintaining such decisions after evidence surfaces that, for instance, *Homo sapiens* and Neanderthals bred healthy, fertile offspring (which is a common, if not universally recognized, indicator that we have in view two races or breeds of a single species rather than two separate species).

Once we become aware of how many times ornithologists decide to reclassify species—merge them, multiply them—based on the latest field observations of a few birds, we begin to appreciate that there is a lot of social construction to the notion of species. We can happily accept the possibility that many of the members of the genus *Homo* are fully human descendants of Adam and Eve.

What about the late flowering of human culture? Here the discussion gets more speculative, with some suggestions rather than definitive answers. One approach to the puzzle is to say that God created an animal with all the physical features of human beings but without the spiritual soul that enables thought at a rational level. I see two problems with such a scenario. The first problem is physical. We human beings are slow, weak, and unprotected. We would be an easy target for hungry carnivores. Surely, such ancestors would have been wiped out by natural selection in no time. The other problem with this scenario is ethical. It would give further ammunition to the culture of death to euthanize those deemed sub-human because their mental abilities are not up to scratch.

Another way to try to reconcile the fossils and the cultural flowering is to insist that the dating is all wrong. Young-earth creationists (YEC) speak of the universe being less than ten thousand years old. But to my lights, taking this route requires turning one's back on too many lines of scientific evidence that hang together very well. Moreover, I am convinced that a faithful reading of Scripture allows for an old earth and universe. Additionally, even without going the YEC route, it could be that our current methods of dating, say, cave drawings from the upper and middle Paleolithic, are way off. Perhaps one or more of these are much older than supposed. The scholarly club of experts of cave art is quite small and it is hard for a layman to get a sense of how much confidence to put into the published dating. It wouldn't surprise me if our dating of these things is so theory-laden as to be far less secure than is regularly asserted.

I am emboldened to offer this hypothesis based on the experience of Giuseppe Sermonti at a meeting on the evolution of primates sponsored by the Pontifical Academy of Sciences in 1982. Sermonti writes:

> The primate paleontologists had to take into account an announcement made a few years earlier by the molecular biologists. The latter had just begun looking into human molecules and those of our cousins, the apes. In the light of certain comparisons and calculations, which it is not my purpose here to illustrate, they advanced the hypothesis that humans and apes had taken their separate directions on the evolutionary tree not more than 1.3 million years ago. This was hard for the paleontologists to swallow, because they had been unearthing human fossils at depths dated at three million years and had found traces of hominids of erect posture (*Australopithecines*) five to six million years old. Since it was received wisdom that humans were descended from the apes, how could there have been humans and hominids antedating by millions of years the fateful split that supposedly gave rise to them? At that time, the spotlight was still on an Asiatic specimen (*Ramapithecus*) that most agreed was a hominid, belonging somewhere on the line leading from apes to humans, yet the impertinent *Ramapithecus* had lived fifteen million years before the split.
>
> Anyone looking at science from outside thinks that scientists are at pains to come up with theories that fit their data. Yet the contrary is often the case (and here we have one example), with scientists striving to force the facts into some preconceived theory. Man the progeny of the ape was gospel for the evolutionists.... Everyone took it as tacitly agreed—as if there were a moral obligation to do so, a categorical imperative never to question it.
>
> In this way a scandalous compromise was reached. The unfortunate *Ramapithecus* was ejected from the hominid branches of our tree and relegated to another tree in the forest. The molecular

biologists agreed to push back the split between hominids to seven million years ago, and the paleontologists promised not to place the hominids at a date earlier than the agreed upon-upon seven million years. The compromise entailed a highly dangerous concession from the biochemists, who thereby had to admit that the so-called "molecular clock" that marked the accumulations of mutations over time had ticked away at a slower rate in the hominid line.[90]

Perhaps I should mention that the Pontifical Academy of Science chooses its members on scientific merit or prestige, not on religious belief. So it is not strange that this august body should collude in favor of human evolution with no regard for Catholic dogma.[91]

As for the paucity of cultural artifacts prior to a few ten thousand years ago, rather than insist on the possible error of dating known cultural artifacts, I find it more promising to propose that most of the older cultural artifacts are lost to the ravages of time. It is reasonable to assume that there were long stretches of time when the artifacts produced by early humans simply disintegrated. We talk about the Stone Age and of cavemen, but think about it: Artifacts are much more likely to last in a cave, tucked away from the weather, than out in the open. And stones preserve much better than wood or bones small enough to serve as hand tools. Stone certainly holds up better than clay or clothes.

We have a clear instance of this in North America. Archaeologists now think that humans crossed into the Americas perhaps as early as 25 Kya. But there is not much evidence of their presence in the north of the continent except for spear and arrow tips and stones arranged in a circular fashion to hold down tents. One might also mention the extinctions of animals such as the mastodon, the camelops, the American lion, the California Turkey, and many others around 12 Kya as evidence of humans moving in. But although this may be a sign of human hunting ingenuity and probably greatly increased numbers of humans on the continent, it does not directly witness to human cultural flowering in the same way that the Chauvet cave paintings do.

We should also question the assumption that humans made in the image of God, with the capacity for language and reason, will necessarily leap forward technologically. Cultures proceed through different phases of development and decay. There is a great difference, for example, between Renaissance Florence and the Fuegian culture Darwin encountered during his voyage on the Beagle. But both the Florentines and the Fuegians were fully human, though clearly in different states of cultural development. If most humans lived in a Fuegian state throughout our long existence, we should not be surprised at the lack of cultural artifacts dating back one to two million years.

Also, was it perhaps far more difficult to accumulate new technological advances when there were as yet no technological advances to build on? In such a primitive state, humans well might face an even more intense struggle for survival than have the primitive hunter-gatherers of a later time, who were armed with fire-starting skills and the bow and arrow. At the beginning, and after the fall of man, the lack of any aiding technologies might have necessitated humans living at even lower population densities than Neolithic hunter-gatherers, making the transmission of new ideas all the more difficult. And if early humans were, as Genesis suggests, particularly prone to violence, this too could have played a role in stalling cultural development.

Another possible reason for the lack of clear human artifacts from much before 50 Kya might be the arrival of the ice ages. We currently are in what is known as the Quaternary glaciation period, which began 2.58 Mya and is characterized by a series of ice ages. During these times, significant parts of Eurasia and North America were covered in ice sheets several kilometers thick. In many places, these sheets scraped away everything to the bedrock. And the subsequent warming and melting did not help to preserve anything that might not have been destroyed.

There is the further important consideration of sea levels. During the Quaternary, glaciers depleted water in the oceans to such an extent that sea levels dropped by up to 120 meters,[92] meaning it was possible to walk from France to England. Throughout most of the last 2.58 million years, ocean levels have been dozens of meters lower than they are today. Thus, when *Homo erectus* arrived on the scene roughly two million years ago, land that is presently deep under seawater was instead oceanic shoreline. Sea coasts provide abundant food resources, and low-lying areas close to the ocean tend to be warmer and more temperate, so they would be attractive places to live. Thus, it may be that many human artifacts from this period have been buried, or have decayed and even dissolved, under the sea.[93]

Other popular places for humans to live, both primitive and modern, are beside lakes and rivers, providing as they do a source of both food and fresh water. But lakes and rivers flood. They rise and fall as climatic conditions change, meaning that the remains of human communities from a million or more years ago beside lakes and rivers likely have been inundated countless times, making artifact preservation over hundreds of thousands of years even more difficult than it already is.

It would, of course, be much easier for Christians if the first appearance of human-like fossils coincided with the first appearance of impressive human cultural artifacts. It would also be easier if the book of Genesis were more transparent to the modern reader regarding the details of early man. But alas, it is not so. Nevertheless, the Church's teachings about human origins do not contradict the evidence of archaeology. And it is only the dogmatic belief in materialism that prevents most modern scientists from recognizing humans as exceptional, different in kind, not just in degree, from all other forms of life. It is perfectly right and just to see in us, although tainted by original sin, the *imago Dei*.

7. Man: The Image of God

When I look at thy heavens, the work of thy fingers,
the moon and the stars which thou hast established;
what is man that thou art mindful of him,
and the son of man that thou dost care for him?

Yet thou hast made him little less than God,
and dost crown him with glory and honor.
Thou hast given him dominion over the works of thy hands;
thou hast put all things under his feet,
all sheep and oxen,
and also the beasts of the field,
the birds of the air, and the fish of the sea,
whatever passes along the paths of the sea.
—Psalm 8

This book revolves around a scientific examination of the physical evidence in nature for intelligent design. But as I made clear in the introduction, the book also sets out to situate questions about biological and human origins in a wider philosophical and theological context. We continue with the latter focus in this chapter.

The biblical view of man and creation conveyed in Psalm 8 above is at odds with the "scientific" conception of humanity's origin. According to the prevailing teaching, we were descended from something that looked like an ape. Just-so stories abound as to why we are how we are. We are bipedal, unlike other primates, because we needed to see our predators from afar on the savanna. We developed the capacity for language and speech because, well, it would certainly help our hunting efforts and binds us together socially. We are the product of a blind tooth-and-claw struggle to reproduce and pass on our genes. We groped our way to self-consciousness. And now we are discovering all sorts of mysteries through science, including the mystery of our own origins.

We are clearly the most clever and adaptable animals on earth. But in the eyes of many, we are a blight or a cancer. After all, it's pointed out, our species has caused deforestation, pollution, massive species extinctions, and global climate change. And now, thanks to the development of nuclear weapons, we may even be on the verge of destroying ourselves through warfare. The Darwinians have just-so stories to explain these darker elements as well. The peaceful and contemplative ape-like predecessors were eliminated by natural selection, which favored the aggressive, the lustful, the hoarders, the murderers.

In Chapter 2 we considered the scientific evidence for a strictly Darwinian (and materialist) origin and diversification of life into the myriad species on earth, and found that evidence inadequate. In Chapter 6, we looked at the evidence for the gradual evolution of man from some ape-like ancestor. It is clear that the fossil record shows no such thing. And the evidence on this point from molecular biology is, at best, a mixed bag and, at worst, weighs decisively against it.

But even if we throw Darwinism overboard and accept the evidence for intelligent design, a puzzle remains. The testimony of both the fossils and genetics point to the arrival of *Homo erectus* as much as 2 million years ago, and yet this predates the evidence for the cultural flowering of prehistoric man by over a million years. To put the matter bluntly, if Adam and Eve were created roughly 2 million years ago, why the long delay?

In this chapter, we will look at further considerations about the origin of man, which the Darwinians do not trouble to ponder, either because these mysteries have no place in their materialist worldview or because they think that one or more of their just-so stories have explained them away. But these mysteries are of fundamental importance to the Catholic understanding of man.

The Bible begins with an account of the creation and fall of man. It provides the broad framework for the understanding of everything else in our faith. It is our "creation myth," provided the term "myth" admits the possibility of truth. The Darwinian account, on the other hand, is the creation myth of atheists.

I am, of course, aware that some sincere Christians have embraced Darwinism and insist that it is compatible with their faith. I will discuss that position in the next chapter. Be that as it may, it is clear that atheists almost unanimously look to Darwinism as the true account of our human origins. That is why countless schoolchildren were marched before Piltdown man at the British Museum until it was finally proved that this supposed fossil of the "missing link" was a forgery. And it is why the education establishment at all levels closes ranks to cancel anyone who calls into question the myth of unguided microbe-to-man evolution.

Happily, it is difficult to cancel the pope.

Pope Saint John Paul II, in his allocution to the Pontifical Academy of Science, said, "The theories of evolution which, because of the philosophies which inspire them, regard the spirit either as emerging from the forces of living matter, or as a simple epiphenomenon of that matter, are incompatible with the truth about man."[1] In other words, Darwinism cannot account for the origin of man, or at least cannot be the full story. In this chapter, we will look at the traits that are specifically human and that cannot be simple epiphenomena of matter, which is to say that they do not arise from matter or its configuration but come from above or beyond the material.

In the Image of the Trinity

In the opening chapter of the book of Genesis, we read, "God created man in his own image, in the image of God he created him; male and female he created them." The previous verse says, "Let us make man in our image, after our likeness." The plural "us" / "our" is puzzling. It could be read as a case of the royal "we," but Christian commentators from early on came to understand it as prefiguring the Trinity. Saint Augustine, for example, developed the psychological analogy of the Trinity based on his understanding of the human soul.[2] The human mind can know and choose. So the Son is likened to the human intellect and the Holy Spirit to the will. But it is really the other way around. We can think and choose because we are made in the image of the Trinity. And our own self-understanding provides us with only the vaguest ideas about the Triune Godhead.

Our being created in the image of the Trinity bestows a great dignity on us. We were made by God as the pinnacle of creation, to have dominion over all the other creatures. In the first chapter of Genesis, we learn that Adam and Eve were made on the sixth day. In the creation account from the second chapter, Adam was made on what appears to be the first day (though the particular day in this account is left unclear) and it was up to him to name the animals as God brought them before him. One can try to harmonize these two creation accounts, but that would take us far afield. The essential point is that both accounts present man as the pinnacle of creation.

The ancients, both Christians and pagan, spoke of man as a microcosm. We share in both the material and the spiritual. There is something of the animal in us as well as of the angelic. All of creation is brought together in us. This makes us exceptional. In terms of how we differ from other animals, Saint Thomas focused on the intellect and the will. The intellect, although mysterious in many ways, is much easier to understand than the will. So let us begin with that.

The Intellect

Saint Thomas teaches that the human soul must be a spiritual reality, meaning that it is not made of matter. He arrives at this conclusion by examining how we come to know universals. Other animals can know particulars: this or that tree they see before them. But only humans can form an idea of a tree in general. If we are to understand what a tree is, we must grasp in our mind the form of a tree. For Saint Thomas, these forms do not exist in some Platonic realm; they exist first in the mind of God and then in individual trees. They also come to exist in our intellect, once we have abstracted the universal form of *tree* from the one or more individual trees we have encountered.

If our idea of *tree* is correct, then the abstract universal idea of *tree* in our minds will correspond with the tree we encounter in the park. If the soul were material, then it could not sustain the myriad different forms we have come to know. The reason for this arises from Aristotelian hylomorphism. According to this view, form (*morphē*) united to matter (*hylē*) is what makes the individual. So the form *tree* united to matter is the tree that we can see and touch. If the mind were something material, the form of the tree could not inhere in it without turning the mind into a tree. Thus the mind could not know many things. It could not be at the same time a tree, a frog, a planet, etc. Were the form in us in any way material, there would not be a correspondence of ideas between our intellect and empirically accessible reality. True knowledge would be impossible.

All sorts of objections might arise at this point. The Platonists would argue for a separate realm of forms. But they would agree that the soul is not made of matter. Nominalists would object that there are no such things as universals. And there would be a loud chorus from academia eager to point out that the coherence theory of truth—favored by social constructivists—does away with the need for ideas to correspond to reality. I cannot resolve the debate in a few paragraphs. But the sympathetic reader should at least be able to see problems with nominalists or social constructivists pontificating about the way that the cosmos really is. All their arguments are basically variations on the self-refuting assertion, "There is no such thing as truth." If that statement were true, it would mean the statement itself isn't true.

It is much easier to point out the deficiencies in nominalism and social constructivism than to establish the correspondence theory of truth. Many propositions that were thought to be demonstrably true in the past have been shown to be false as various sciences progress. Saint Thomas, for example, chose to illustrate what he meant by a demonstrated truth by appealing to physics which, since at least the time of Plato, had taught that "the movement of the heavens is always of uniform velocity."[3] Clearly it is not.

Saint Thomas's argument for the immateriality of the intellect might be too technical for most people, especially if they are not trained in scholastic philosophy. So it may be useful to provide a more commonsense approach. This approach perhaps makes it clearer why grounding our ability to think in matter undermines any confidence we naturally have that we can know truth about an objectively real world. Common sense tells us that there are things that we can know. And most people, other than those brainwashed by the prejudices of academia, have a hard time believing that the ability to know, say, basic truths of logic or math or observation, comes from mere matter as known by physicists and chemists. If at bottom we are nothing but chemical soup, how can we possibly think we can come to real knowledge of anything?

Darwin himself worried about this. In a letter to William Graham, he wrote, "With me the horrid doubt always arises whether the convictions of man's mind, which has been developed from the mind of lower animals, are of any value or at all trustworthy. Would anyone trust in the convictions of a monkey's mind, if there are any convictions in such a mind?"[4] Whereas Darwin chose to suppress his doubt and refused to entertain any exception for the human mind in his materialistic origins story, even some of those who most sympathized with natural selection did not follow him in his denial of a spiritual mind.

Asa Gray, for example, wrote a defense of the power of natural selection against criticisms made by his fellow American scientist Louis Agassiz; but in doing so, Gray just assumed that Darwin could not possibly have meant his theory to explain the human mind as just a refined form of animal instinct. Darwin politely objected to Gray in his private correspondence, and yet, as Benjamin Wiker points out, Darwin "had no qualms about using Gray's arguments if it would smooth the way for the acceptance of his theory. Once the theory was accepted, the theistic patina [which Gray had put on it] would be ground away by the hard, anti-theistic core of the argument."[5]

Darwin also felt "deeply disappointed"[6] when geologist Charles Lyell refused to believe that man was part of the evolutionary continuum. Darwin would face other such disappointments. The co-discoverer of the theory of evolution by natural selection, Alfred Russel Wallace, also came to regard their evolutionary mechanism as more limited in reach than Darwin wished. "Neither natural selection," wrote Wallace, "nor the more general theory of evolution can give any account whatever of the origin of sensational or conscious life."[7]

It is a tribute to Darwinian propaganda that the great majority of academics today think that chemicals can give rise to conscious thought and true knowledge. Some, such as Evelyn Fox Keller, at least have the clear-headedness to acknowledge that if natural selection is all there is to our development, then we will never know many things, because natural selection could not have honed our minds sufficiently.[8] Much better is Einstein's sense of wonder that the human mind can come to know the cosmos. And what best grounds that reliable intuition, effectively explaining both how we know and why we can confidently believe that we will come to know still more about the cosmos in the future, is the idea that each of us possesses an immaterial soul, given to us by a Creator who formed the universe according to a rational plan.

So far, this has been a general discussion about the human ability to know. It has been about the human intellect. One source of doubt that the human mind is exceptional is the various levels of intelligence found in animals, and in devices created by humans. There are people who speak of the intelligence of certain plants, such as sunflowers in tracking the sun, or of the intelligence of salmon in navigating back to the gravel beds in which they began their life. Bees display intelligence in building their hives and in distinguishing between members of their hive and others. The more biologists investigate, the more wonders of animal intelligence they discover. In some cases, this native intelligence goes clearly beyond what one might call instinct, as in the case of animals that can be trained to do some fairly complicated things: rats learning their way out of a maze, or dogs, seals, and chimpanzees trained to perform impressive tricks.

As for machines, we are all familiar with various systems of artificial intelligence: spell checkers, grammar checkers, translators, face and speech recognition, autopilots in aircraft and self-driving cars, formidable chess programs, and chatbots. Those who would like to deny human exceptionalism point to all these various levels of intelligence and insist that there is no real difference in kind, that it is just a matter of degree.

However, several different considerations, taken together, strongly suggest that the mind is an immaterial reality. Much like the multiple ways that lead to the certain conclusion that God exists, arguments for the spiritual nature of the soul function together as "converging and convincing arguments, which allow us to attain certainty about the truth."[9]

The Origin of Human Language

A very biblical place to start in our exploration of the human intellect is with human language. The Gospel according to John begins, "In the beginning was the Word (*logos*)." The Word that became flesh is the second Person of the Holy Trinity, He who died for the sins of the world. The Bible also calls Him the Wisdom of God. The connection between language and wisdom goes back a long way, for the classical Greek word "logos" means both the "word by which the inward thought is expressed" and "the inward thought" or "reason itself."[10]

No other animal besides man has the ability to speak a language. Yes, animals can communicate, either by sounds or various bodily gestures. But we never hear them or see them communicating in a manner that would begin to resemble the complexity of human speech. In the early 1960s Noam Chomsky convinced most people that all human languages share an invariant universal grammar. The various languages might be different in their specific grammars, but these specific grammars never go outside the limits of the universal grammar.

This ability to acquire and use language, Chomsky thought, was a quirk of evolution. He believed that there was a physical speech organ, but he was rather vague about it—"specific neural structures, though their nature is not well understood."[11] Speech, according to Chomsky, is a quirk of evolution because there was no selective pressure to develop this ability. It is restricted to human beings. And if one accepts an ancient history for the genus *Homo*, it is truly a cause for amazement that people whose last common ancestor might have lived over a million years ago, should be able to learn each other's languages relatively easily—especially if they are children. As Chomsky wrote:

> In New Guinea, Australia, and so on… there are… what we call "primitive people" who to all intents and purposes are identical to us. There's no cognitively significant genetic difference anyone can tell. If they happened to be here, they would become one of us, and they would speak English; if we were there, we would speak their languages. So far as anyone knows, there is virtually no detectable genetic difference across the species that is language-related…[12]

Chomsky's claims have, of course, been challenged. In particular, in the early 1970s, Herbert Terrace, a psychologist at Columbia University, tried to teach a young chimpanzee, named Nim Chimpsky, to communicate using American Sign Language. Although Nim learned to recognize about 125 signs, Terrace doubted that the animal had acquired anything that shared the traits of human language. He conceded that Chomsky was, in fact, right.[13]

Other researchers are still trying to coax out some language abilities from animals. And, insofar as animals can communicate in nature and be trained by humans, there will always be some evidence to suggest that human language is just a more complex version of whatever it is that animals have. But as it stands, most of us recognize that human language is in a league by itself. It is something different in kind, not just degree, from animal communications.

A more serious challenge to Chomsky's theory came from Daniel Everett's study of the Pirahã tribe, who live deep in the Amazon rainforest of Brazil. Apparently the tribe's language does not, at least in one crucial respect, follow the rules of universal grammar. Tom Wolfe, in *The Kingdom of Speech*, recounts a decade-long battle between Everett, on one side, and Chomsky and his supporters, on the other. On Everett's understanding, language does not depend on a biological structure. It is an artifact. But that hardly solves the problem as to why only human beings have developed this artifact. What is it about us that makes us capable of understanding and using it?

Some people would like to attribute our language-speaking ability to genetics. In Chapter 6 we encountered the gene *FOXP2*, which is different in the genus *Homo* from what is found in chimpanzees and other animals. Moreover, mutations in *FOXP2* in modern humans cause speech disorders. But, as Michael Denton points out, the evidence is controversial, because there is evidence that the gene has influence on many different aspects of brain development.

It is clear that there is a material component to our ability to speak—besides the obvious need for vocal cords, tongue, lips, and teeth—because brain lesions cause various aphasias (the loss of previously possessed linguistic abilities). Probing the brain can do the same. But it is interesting, as Michael Denton points out, that "no genetic mutations have ever been observed to cause the sorts of specific defects in brain processing which follow from brain lesions in the adult brain." Denton uses this to argue that the organization of the adult brain cannot be genetically specified but must "arise entirely from higher organizational phenomena during development."[14] He does not specify how this self-organization occurs, but he is confident that it did not arise from Darwinian selection.

To add to the mystery, Denton cites the work on neuroplasticity by Norman Doidge. Doidge reports that a woman born with half a brain was able to fool neurologists into thinking that there was nothing wrong with her.[15] There are other examples. In the 1980s, John Lorber, a professor of pediatrics at Sheffield University, came across a mathematics student with an IQ of 126 who had almost no brain. He suffered from hydrocephalus as an infant and, as a result, his layer of cerebral tissue was only one millimeter thick, instead of some 4.5 centimeters as it is in an average adult.[16] No one has yet been discovered alive and functioning with no brain, so it is evident that the brain is necessary. But it is not at all clear how thought is related to it.

Neuroplasticity, the ability of the brain to adapt to lesions and diseases, can point beyond a genetic or mechanical organization of the brain and its functioning. This phenomenon is not, however, limited to humans. Denton cites an experiment at MIT in 2000 in which the researchers "rerouted the visual input from the retina to the auditory cortex in newborn ferrets." The ferrets were able to see with their auditory cortex.[17] Clearly something is presiding over the reorganization of the brain, and it does not appear to be genetic. Neuroplasticity, it seems to me, is a good argument for the Aristotelian conception of the soul as the principle of life. That brings us one step closer to the human mind, which is a faculty of the soul. But, by itself, it does not establish that the human soul is something different in kind from an animal soul.

The evidence for human exceptionalism is, of course, plainest at the behavioral level. For instance, only human beings create art. If we see a cave painting, no matter how old it is, we know that it was made by humans rather than by, say, chimps or raccoons. Other animals might use objects as tools or even make tools—if smashing a rock to make a sharp edge counts as tool-making—but only human beings decorate their tools. In Chapter 6 we considered carvings (the Venus statuettes), jewelry, and musical instruments. Some of the representational art in the prehistoric caves is of the highest quality. It gives empirical support to Chomsky's claim that "what we 'call primitive people'... to all intents and purposes are identical to us."[18]

Whether the art is representational or more abstract, such as music, it gives an indication of an inner life of the human mind that is radically different from animal consciousness. Man is clearly self-conscious—witness the pictures of hunters chasing game. And he desires to find some meaning to his existence. In fact, Rodney Stark, who studied the history of religion, maintained that there is good reason to believe that "primitive" societies had conceptions of a High God.[19]

In light of all this, we can go back to language and look at an interesting argument made by Louis de Bonald in the beginning of the nineteenth century. De Bonald was writing in the aftermath of the horrors of the French Revolution, which, he argued, were the necessary consequences of mankind's turning its back on the institutions that God had given to our first parents. This will not get him a sympathetic hearing in today's world. He can also be slammed for being anti-Semitic. Nevertheless, he provides us with an important consideration about the origin of language.

Language, de Bonald noted, requires thought; but thought—or at least most of what we take as thought—requires language. Without being able to think, one could not develop language. But one cannot think much beyond the early cognitive experiences of a newborn without a language. It is no surprise that "logos" can mean both word and thought itself. De Bonald maintained that language was given to our first parents along with the ability to think.[20] Children can now pick up a language by natural means. But that is only because their elders have the gift of language for children to mimic.

What of the competing explanation offered by Darwinism? Denton quotes David Premack to give us a sense of how unlikely it is that language should have evolved through natural selection:

> I challenge the reader to reconstruct the scenario that would confer selective fitness on recursiveness. Language evolved, it is conjectured, at a time when humans or protohumans were hunting mastodons... Would it be a great advantage for one of our ancestors squatting alongside the embers, to be able to remark, "Beware of the short beast whose front hoof Bob cracked when, having forgotten his own spear back at camp, he got in a glancing blow with the dull spear he borrowed from Jack"?
>
> Human language is an embarrassment for evolutionary theory because it is vastly more powerful than one can account for in terms of selective fitness. A semantic language with simple mapping rules, of a kind one might suppose that the chimpanzees would have, appears to confer all the advantages one normally associates with discussions of mastodon hunting or the like. For discussions of that kind, syntactical classes, structure-dependent rules, recursion and the rest, are overly powerful devices, absurdly so.[21]

Not surprisingly, Chomsky is also skeptical of the Darwinist claim: "If you look at the literature on the evolution of language, it's all about how language could have evolved from gestures, or from throwing, or something like chewing, or whatever. None of which makes any sense."[22]

By 2014, Chomsky, somewhat chastened by his prolonged clash with Everett, nevertheless added his name to that of seven other luminaries in the study of language, in the article "The Mystery of Language Evolution." Note the word "mystery." There we find this pointed comment: "In the last 40 years, there has been an explosion of research on this problem as well as a sense that considerable progress has been made. We argue instead that the richness of ideas is accompanied by a poverty of evidence, with essentially no explanation of how and why our linguistic computations and representations evolved."[23] Language was a stumbling block to Darwin right from the beginning. Wallace used it to argue for human exceptionalism. And that is where the argument remains today.

Evidence from Neuroscience

Another line of evidence for the spiritual nature of the mind is the work of the neurosurgeon Wilder Penfield. In the 1930s, he became the first director of the Montreal Neurological Institute and Hospital, where he performed brain surgery on patients under local anesthesia. The goal of these procedures was to cure, or at least to mitigate, the patients' epilepsy. But, in the process, he learned a lot about the various areas of the brain and their function.

A classic one-minute video produced by Historica Canada summarizes his work on epilepsy and ends by calling him the greatest Canadian alive.[24] The video begins with a distressed woman telling her husband that she smells burnt toast. He is about to tell her that he cannot smell it, but before the words are out of his mouth, she drops the tray she was carrying in a fit of epilepsy. We next see her in the operating room. Her brain is exposed and Dr. Penfield is stimulating it with his electrodes. She first sees the most wonderful light. Then she thinks that Penfield has poured cold water on her hands. Finally, she smells burnt toast. The video does not show the operation removing part of the brain. It ends with her thanking Dr. Penfield for curing her and hundreds of other patients of epilepsy.

Another neurosurgeon, Michael Egnor, tells us that Penfield came to the field of neuroscience as a convinced materialist, viewing the mind as completely explicable by the brain. But he changed his mind in light of what he himself discovered. Penfield noted that stimulating the brain with a low voltage could evoke sensations and various perceptions, memories, or even emotions, and sometimes it could cause a seizure or the movement of muscle. But it never evoked abstract thought. As Penfield put it:

> There is no area of gray matter, as far as my experience goes, in which local epileptic discharge brings to pass what could be called a "mindaction"... there is no valid evidence that either epileptic discharge or electrical stimulation can activate the mind... if one stops to consider it, this is an arresting fact. The record of consciousness can be set in motion, complicated though it is, by the electrode or by epileptic discharge. An illusion of interpretation can be produced in the same way. But none of the actions we attribute to the mind has been initiated by electrode stimulation or epileptic discharge. If there were a mechanism in the brain that could do what the mind does, one might expect that the mechanism would betray its presence in a convincing manner by some better evidence of epileptic or electrode activations.[25]

Penfield concluded that since abstract thought could be totally switched off by drugs or brain stimulation, the mind was dependent on the brain. But since it could not be turned on in a physical way, abstract thought was something more than physical.

Egnor, in his article on Penfield, cites another telling passage:

> The patient's mind, which is considering the situation in such an aloof and critical manner, can only be something quite apart from neuronal reflex action. It is noteworthy that two streams of consciousness are flowing, the one driven by the environment, the other by an electrode delivering sixty pulses per second to the cortex. The fact that there should be no confusion in the conscious state suggests that, although the content of consciousness depends in large measure on neural activity, awareness itself does not.[26]

Egnor notes that Penfield finished his career as someone who believed that the mind was an immaterial reality. Egnor holds this position himself, but reminds his reader that it is hardly novel. Aristotle came to that conclusion over two thousand years ago, for, as Egnor puts it, "universals—concepts that are not particular things—by their nature cannot be in particular things, and thus cannot be in matter, even in brain matter."[27]

Similarly, the "modern" idea that thought is just a material process goes back at least two millennia. In the book of Wisdom, the wicked are depicted as viewing their life and their thoughts thus: "Short and sorrowful is our life, and there is no remedy when a man comes to his end, and no one has been known to return from Hades. Because we were born by mere chance, and hereafter we shall be as though we had never been; because the breath in our nostrils is smoke, and reason is a spark kindled by the beating of our hearts" (Wisdom 2:1–2).

So we see that the debate about how thought is possible, and how it relates to matter, has been going on for quite some time.

Egnor is not the only dualist neuroscientist after Penfield. One can mention the Nobel prize winner John Eccles. In a book he wrote with the psychologist Daniel Robinson, the authors state: "We regard promissory materialism as superstition without a rational foundation. The more we discover about the brain, the more clearly do we distinguish between the brain events and the mental phenomena, and the more wonderful do both the brain events and the mental phenomena become. Promissory materialism is simply a religious belief held by dogmatic materialists... who often confuse their religion with their science."[28] Another dualist, Mario Beauregard, is co-author with Denyse O'Leary of *The Spiritual Brain: A Neuroscientist's Case for the Existence of the Soul*.

Even neuroscientists who are not inclined to accept a spiritual account of the mind are admitting that their field of knowledge is not getting anywhere. Thus, for example, British neuroscientist Matthew Cobb writes:

> And yet there is a growing conviction among some neuroscientists that our future path [to understanding how the brain works] is not clear. It is hard to see where we should be going, apart from simply collecting more data or counting on the latest exciting experimental approach. As the German neuroscientist Olaf Sporns

has put it: "Neuroscience still largely lacks organising principles or a theoretical framework for converting brain data into fundamental knowledge and understanding." Despite the vast number of facts being accumulated, our understanding of the brain appears to be approaching an impasse.[29]

The spiritual soul is far from being disproved by materialist neuroscience. Rather, the field provides much evidence for it.

The Undiscovered Country… Discovered?

Another body of evidence highlighting the difference between the brain and the mind is that of near-death experiences (NDE). NDEs, as their name suggests, are the experiences of those who have had close brushes with death, and yet survived the experience to talk about it. In these cases the patients lost normal consciousness—some were declared clinically dead—and yet they described their experiences from that period as being more real than dreams and sometimes even more real than the world that we all inhabit. Although such cases may have been around since prehistoric times—and some researchers even suggest that they may have given rise to the popular artistic images of angels—it is only with the rise of modern medicine that the number of cases has skyrocketed. That should not be surprising, because modern life-support systems have made it possible for people to recover from trauma or disease that would have killed them in earlier times.

At first, the medical profession tended to downplay these experiences as hallucinations caused by a distressed brain. It was also possible to cast doubts on the veracity of the person describing the experience. There are several bestselling books written by those who claim to have experienced an NDE, so there is an obvious financial incentive to make up stories. But the large number of cases and some of the details mentioned by these NDErs have finally succeeded in attracting the attention of mainstream publications such as *Psychology Today*. There Steven Taylor wrote:

> NDEs have never been satisfactorily explained in neurobiological terms. Various theories have been suggested, such as hallucinations

caused by a lack of oxygen to the brain, undetected brain activity during the period when the brain appears not to be functioning, the release of endorphins, a psychological 'depersonalisation' in response to intense stress, and so on. All of these theories have been found to be problematic.[30]

It would take us too far afield to discuss the various NDEs in detail, but a few comments should indicate their relevance to the discussion of the nature of the mind. First, according to one skeptical researcher, Christof Koch, both those who are religious and those who claimed no prior religious belief are well-represented among those who reported NDEs.[31] This tends to rule out that the brain is drawing on prior prejudices and desires.

In fact, NDEs can be problematic for devout Catholics if they are understood as real beginnings of the journey to heaven which got derailed through the machinations of doctors. The great majority of reported NDEs—upwards of 85 percent—are intensely affirming, peaceful, and pleasurable. One would think that most people would require purgation of their sinful attachments before enjoying the bliss of heaven. Nevertheless, there are some horrific examples of NDEs as well, suggesting descents into hell or perhaps purgatory. It has also been suggested, with some evidence, that people who experience horrific NDEs may be likely to suppress them, meaning that the percentage of horrific NDEs is much higher than remembered and reported.[32]

If one sees the NDE as an enticement or a warning or just an expansion of one's vision of the world, there need not be anything contrary to the Christian faith in them. They could be special acts of God's providence to lead people to salvation, either by offering a carrot or warning with a stick. And, in fact, those who have had NDEs tend to become more spiritually inclined and less fearful of death. Mere hallucinations do not tend to have such permanent effects on people.

A common feature of many NDEs is that the person has a sense of watching himself from a distance, such as the ceiling of an operating room. It is not just a matter of them experiencing the scene from a different perspective, for in many of these cases the eyes of the patient are taped shut so as to prevent injury from the bright lights in the operating room. The details that such people provide are of great interest because they can be verified by the doctors and nurses and often they are exactly correct.

One famous case involved Vicki Umipeg of Seattle, who had been born blind. She was 21 years old in 1973 when she was critically injured in a car accident and was clinically dead for about four minutes. That is, her heart had stopped and she had lost consciousness. She describes herself as looking down from the ceiling and seeing herself being worked on. She was terrified at the vision because she had never experienced sight, but quickly adjusted to the situation. In particular, she mentioned that the woman on the emergency room table had a ring on the same finger on which she usually wears her ring, so she recognized her to be herself. She was also able to report the conversations of the doctors. After she was resuscitated, she remained blind.[33] This "disconnect" of the mind from the body is certainly a challenge to a purely materialist explanatory scheme.

Another often reported feature is that people see deceased loved ones during their NDEs. Pim van Lommel was a Dutch physician specializing in NDEs. He reports the case of a five-year-old girl who had fallen into a coma through meningitis. In her experience, she saw a girl who was about ten years old. The girls embraced and the older girl introduced herself as her sister, Reitje, named for their grandmother. When the five-year-old woke from the coma, she related the story to her parents, who panicked and left the room. They came back later and admitted that they had a girl named Reitje who had died earlier of poisoning. Up until then, they had kept it a secret, to be revealed when their children could begin to understand.[34]

Multiple books could be filled with similarly well-attested cases in which the hallucination explanation fails to account for all the facts. Will such evidence convince the committed materialist that the mind is a non-material reality? Probably not. The philosopher Neal Grossman reports a conversation he had about NDEs with one such type.

> Exasperated, I asked, "What will it take, short of having a near-death experience yourself, to convince you that it's real?"

Very nonchalantly, without batting an eye, the response was: "Even if I were to have a near-death experience myself, I would conclude that I was hallucinating, rather than believe that my mind can exist independently of my brain."[35]

That kind of attitude is exactly what Jesus described in the parable of Lazarus and the rich man: "If they do not hear Moses and the prophets, neither will they be convinced if some one should rise from the dead" (Luke 16:31).

Terminal Lucidity

Another line of evidence for the immaterial nature of the mind comes from the widely observed fact that people with severe mental deficiencies sometimes become perfectly clear and sensible a short while before they die. It happens often enough that it has its own term in the field of hospice care: "terminal lucidity." One of the more famous such cases is that of Anna Katharina (Käthe) Ehmer (1896–1922). Friedrich Happich, a Protestant minister in charge of the institution where Anna lived from the age of six until her death, and who had known her for many years, provides this picture of Käthe:

> Käthe was among the patients with the most severe mental disabilities who have ever lived in our institution. From birth on, she was seriously retarded. She had never learned to speak a single word. She stared for hours on a particular spot, then she fidgeted for hours without a break. She gorged her food, fouled herself day and night, uttered an animal-like sound, and slept. In all the time she lived with us, we have never seen that she had taken notice of her environment even for a second. We had to amputate one of her legs, she wasted away.

So, it was quite a surprise to see a great change in her shortly before her death. Happich writes:

> One day I was called by one of our physicians, who is respected both as a scientist and a psychiatrist. He said: "Come immediately to Käthe, she is dying!" When we entered the room together, we did not believe our eyes and ears. Käthe, who had never spoken a

single word, being entirely mentally disabled from birth on, sang dying songs to herself. Specifically, she sang over and over again, "Where does the soul find its home, its peace? Peace, peace, heavenly peace!" For half an hour she sang. Her face, up to then so stultified, was transfigured and spiritualized. Then, she quietly passed away. Like myself and the nurse who had cared for her, the physician had tears in his eyes.[36]

Michael Nahm and Bruce Greyson have carefully examined the evidence for this case and think Happich and the physician mentioned are credible witnesses, although the historical distance makes it impossible to question witnesses.[37] However, there are more recent cases, which Nahm and Greyson report in another paper. Here is one of several:

> Morse and Perry (1990) reported the case of a 5-year-old boy who had been in a coma for three weeks dying from a malignant brain tumor, during which time he was almost constantly surrounded by various family members. Finally, on the advice of their minister, the family told the comatose child that they would miss him but he had their permission to die. Suddenly and unexpectedly, the boy regained consciousness, thanked the family for letting him go, and told them he would be dying soon. He did in fact die the next day.[38]

If the materialist view that the mind is just an aspect of the brain were true, it is not likely that such bouts of mental clarity and verbal capacity would be the result of a brain in the process of shutting down.

The Placebo Effect

Let us now turn to the placebo effect. In testing the efficacy of new medicines, it is common to do a study in which some patients are given the new drug while others are given a pill—a placebo—that looks like the new drug but is in reality something benign and ineffective such as a sugar pill. The great mystery—or problem for a materialist conception of the mind—is that often between 35 percent and 45 percent of the patients who are given the placebo show marked improvements. Some are even completely cured. A new drug, before it is licensed, usually has to be at least 5 percent more effective than the placebo. There is also evidence that placebo surgeries can be effective. Patients have been taken into the operating room and given some slight incisions. Amazingly they report that their knee problems have greatly improved.[39]

Placebo means "I will please." Until the twentieth century, the lion's share of medical treatment consisted of actions now known to be ineffective or even counterproductive in most or all situations—as for example, bleeding to help balance the four humors of the body. If these actions did any good, it must have come about through the placebo effect. The patient is pleased to see the doctor. The doctor, the expert, assures the patient that the action is likely to speed healing. The patient has confidence in the doctor, and this confidence in his mind has a curative effect on his body.

There is also a "nocebo" effect. Beauregard and O'Leary report, among other strange facts, that "volunteers for medical studies who have been warned about the side effects of the medication often develop those effects even though they are in the sugar pill control group."[40]

The placebo and nocebo effects bring up interesting moral questions. Should a doctor warn a patient about the possible bad side effects of a medicine? Should he give a placebo to a hypochondriac so as not to pump potentially dangerous chemicals into his body? We can leave those questions aside here, content to acknowledge that the placebo and nocebo effects are clear examples of a mind working on the body. For those who think that the mind is just an epiphenomenon of brain chemistry, this is at best an embarrassment.

The Turing Test

We now turn to a popular avenue that materialists have been using to argue against a spiritual mind: artificial intelligence (AI). No one thinks that a computer has a soul created by God. So if a computer can be made to fully replicate human intelligence, then it would be clear that the intellect can arise out of matter that is intelligently arranged in a certain way. Note that many devices, even mechanical ones, can be said to mimic some aspect of intelligence. A coin-counter, which separates coins by size, is one such example. So artificial intelligence in a narrow sense is clearly possible. The question is whether general intelligence with conscious understanding is possible. This is known as "strong AI." In contrast, a machine that can only act as though it has a mind and consciousness is usually said to display "weak AI."[41] With the arrival of powerful AI such as ChatGPT, it may seem to many people that the transhumanist dream is just a matter of time and more CPU speed and storage capacity. Nevertheless, many experts in the field are pouring cold water on this view, often in book-length arguments. So this will only be a brief sketch of why it is not reasonable to believe that strong AI can arise from matter without a spiritual component.

Alan Turing was one of the early pioneers of computing. In 1950, he wrote a paper, "Computing Machinery and Intelligence," in which he took up the question of whether computers could think. He was aware of theoretical reasons against it, stemming from Gödel's theorem, to which we will turn shortly. But after the success of using computing machines to help decipher the German Enigma codes, Turing came to believe that the question of human intelligence could be decided empirically, without interminable philosophical debates as to what knowledge and understanding might mean.

Turing proposed a practical test for intelligence, which was a modification of a popular party game called "Imitation." The party game had a man and a woman out of sight in two different rooms. The other guests would be allowed to ask questions of each to decide in which room the woman could be found. Both the man and the woman would try to convince the questioners that they were the woman. Any obvious clues, such as the depth of voice, would be eliminated by limiting the communication to exchanging notes on paper written and delivered by a runner.

In Turing's version, the man and woman are replaced by a computer, which is programmed to trick the questioner into thinking it is human. The communication is restricted to text messaging. In 1990, an annual competition called the Loebner Prize was established to encourage the development of artificial intelligence. The computers did not do well. The biggest splash happened in 2014 when a program named "Eugene Goostman" actually managed to fool one-third of the judges, who were allowed no more than five minutes of questioning. A short snippet of one conversation ran:

> Judge: Why do birds suddenly appear?
>
> Eugene: Just because 2 plus 2 is 5! By the way, what's your occupation? I mean—could you tell me about your work?

George Montañez quoted this in a paper, "Detecting Intelligence: The Turing Test and Other Design Detection Methodologies."[42] Elaborating on this in a podcast, he mentioned that the creator of the program made Eugene Goostman pretend to be a thirteen-year-old Ukrainian boy. That way, the judges, would attribute his lack of general knowledge to his young age, and his grammatical mistakes and non-sequiturs to his age and poor English abilities.[43]

Eugene Goostman's performance was greeted with much media attention. *Gizmodo* informed its readers, "This is big," but it soon changed its mind and published an article claiming, "Turing's test was 'bullsh*t.'"[44] Eric Larson explains the reasons for the about-face in *The Myth of Artificial Intelligence*.[45] People caught on to the fact that "Goostman" was really just a parlor trick, relying on the human experience that many teenagers can be rude and unfocused. Most judges took less than five minutes to figure it out; and no doubt the minority who were taken in would have come to the correct conclusion had the "conversation" been allowed to go on much longer.

But hope springs, if not eternal, then at least for the time being. With each new jump in technology—computer speed, interconnectivity, availability of data—there is some hope that full artificial intelligence is right around the corner. After all, computers can be made to do some pretty amazing—if highly circumscribed—tricks that capture the public imagination.

We are familiar with the ability of programs such as Google Lens to identify pictures. The technology works by feeding massive amounts of data into the program, which develops the ability to identify certain patterns. The curious aspect of these programs isn't when they work as intended by the creators of the underlying algorithm. The curious thing is when the program misfires. Although the programmers know how the algorithms work, they often are at a loss to explain why a particular picture is sometimes misidentified. That would require a step-by-step analysis of all the pictures that the system has analyzed and the various feedbacks it received from users to its prompt "Was this helpful?" To be sure, such programs have very high rates of accuracy, but some of their mistakes clearly reveal that there is no general intelligence or consciousness lurking in the algorithm and data. Otherwise, they would not return "almost certainly a cheetah" from something that looks like pure noise.[46]

The public is also aware of the improvements in machine translations. The programmers of Google Translate have succeeded in getting the program to produce usable translations by having it analyze a vast amount of good human translations of documents available on the web. But the Google translations are far from perfect and occasionally are bizarre in ways that raise eyebrows. In 1964, an initial enthusiast for artificial intelligence, Yehoshua Bar-Hillel, spelled out the problems with machines trying to understand human language. It cannot be done unless the machines really understand what the words mean. He took as his example "The box is in the pen." A human translator would at first think this is strange because boxes do not fit into pens. But then he would remember that "pen" can mean a "fenced-off area" and proceed to do the translation. It is rather strange that, to this day, Google Translate continues to render the phrase into French as "La boîte est dans le stylo."[47] The French word "stylo" refers to a pen in the sense of a writing implement. Why did Google Translate not opt for the French term for a fenced-in area? Because it doesn't deal with meaning. At its root, it's just shuffling around 0s and 1s, following the algorithm given to it by human programmers.

Eric Larson discusses this example as well as other questions that stump computers. Google can quickly answer questions such as "What is the population of Toronto?" or "When was the battle of Waterloo fought?" because such information is found in thousands of web pages. But when such information is missing, it is stumped, even when the answer could easily be reasoned out by a third-grader of average intelligence. Most humans will immediately answer that crocodiles cannot run steeplechases, but what is a computer to do when the question has not been discussed at length, or at all, on the web? Reason its way to an answer? No, because that's not in its toolkit of capacities.

Computers are also stumped by sentences such as "The trophy would not fit in the brown suitcase because it was so small." What was so small? The trophy or the brown suitcase? Larson points out that a trick has been specially devised to help computers answer such a question, but the trick fails if the question is altered ever so slightly to "The trophy would not fit in the brown suitcase despite the fact that it was so small."[48] Again, a child of modest intelligence could figure this out, but a computer trained on more data than is contained in a dozen public libraries is utterly stumped.

The public is aware that computers have excelled at chess, *Jeopardy!*, and Go, defeating the best human champions of these games. Such feats are possible because the tasks are highly circumscribed and amenable to algorithms. But step one foot beyond the programmed boundaries and it immediately becomes clear that they are devoid of understanding.

The Chinese-Room Experiment

It may be helpful to introduce the thought experiment published by the philosopher John Searle in 1980. A non-speaker of Chinese is locked in a room with cabinets full of Chinese documents. He has been given a set of instructions on how to manipulate Chinese characters. Pieces of paper with Chinese characters are slipped under the door to him. He then goes to a filing cabinet, scans through it till he finds the folder matching the Chinese characters he has been handed, copies down the response message from that folder, and slips the response, written in Chinese, back under the door to the questioner outside. The person outside the door receives an apt and coherent response to his question, so naturally he believes that the person inside must understand Chinese, whereas in reality he does not. He doesn't understand the original question. He doesn't understand the answer.

Not surprisingly, this thought experiment, dubbed the Chinese-room experiment, has spawned much discussion, which would take us too far afield to analyze. What is amusing is that the *Stanford Encyclopedia of Philosophy* suggests that one of the points is that "minds must result from biological processes; computers can at best simulate these biological processes."[49] How on earth does this thought experiment tell us that minds must result from biological processes? It tells us nothing of the kind. What it does strongly suggest is that computers, at best, can only simulate a conscious, understanding mind.

Gödel Rains on AI's Parade

Kurt Gödel's theorem poses another argument against the possibility of a computer program's gaining real intelligence and real understanding. John Lucas in the 1960s, and more recently Roger Penrose, developed this argument against strong AI. Any summary of the argument is going to be less than rigorous, but a brief sketch is in order. In the 1930s, Gödel put an end to the logistic project to reduce mathematics to logic. As Stephen Barr nicely summarizes it:

> What Gödel showed, however, and rocked the mathematical world by showing, was that mathematics could not be so mechanized. In particular, he demonstrated that if one is given any consistent formal mathematical system rich enough to include ordinary arithmetic, then there exist propositions (called "Gödel propositions") that (a) can be properly stated or formulated in the symbolic language of that system, (b) cannot be proven using the mechanical symbolic manipulations of that system, and yet (c) can nevertheless be proven to be true—by going outside the system. Because the human mind can grasp the structure of the formal system and the meaning of its symbols, it is able to reason about them in ways that are not codified within that system's rules.

Barr then quotes Penrose:

> One might imagine that it would be possible to list all possible obvious steps of reasoning once and for all, so that from then on everything could be reduced to computation—i.e., the mere mechanical manipulation of these obvious steps. What Gödel's

argument shows is that this is not possible. There is no way of eliminating the need for new "obvious" understandings. Thus, mathematical reasoning cannot be reduced to blind computation.

To jump quickly to the conclusion, since human beings are capable of mathematical reasoning, the brain cannot be reduced to a computer.[50]

The interested reader can refer to Stephen Barr's longer exposition of this point. As one might imagine, many have tried to find some loopholes in Penrose's arguments, attempts which he has shown are without force. The really bizarre point is that Penrose is a materialist. He accepts that the laws of physics as we know them will only allow us to build machines that work computationally and hence will not be able to think as humans. But rather than conclude that the mind is not material, he postulates the existence of new laws of physics that will allow us to build such a machine.

It may not be out of place to mention here that Gödel was a proponent of intelligent design, before the term was coined. In a letter he wrote to Hao Wang in 1972, we find:

> I believe that mechanism in biology is a prejudice of our time which will be disproved.
>
> In this case, one disproof, in my opinion, will consist in a mathematical theorem to the effect that the formation within geological times of a human body by the laws of physics (or any other laws of a similar nature), starting from a random distribution of the elementary particles and the field, is as unlikely as the separation by chance of the atmosphere into its components.[51]

No doubt, despite the proof that human intelligence is not computable, there will continue to be prominent voices telling us otherwise. Whether they look in hope to the future possibility of attaining immortality through an upload of their brain to the cloud, as Ray Kurzweil does, or whether they worry (as Stephen Hawking used to worry) that AI will build better and better AI and either enslave or eradicate humans, their voices will be given credence by the successes of AI. Fortunately, there also will continue to be highly credentialed people who throw cold water on the dream or nightmare.

Penrose has already been mentioned. There is also Gregory Chirikjian, a former director of the Johns Hopkins robotics lab. "Robots will not be able to do skilled work such as carpentry or auto repair or plumbing for the foreseeable future," he insists. "Nor will robots be able to exhibit any form of creativity or sentience."[52] The CEO of Microsoft, Satya Nadella, sounds a similar note. "One of the most coveted human skills is creativity, and this won't change," he writes. "Machines will enrich and augment our creativity, but the human drive to create will remain central."[53] No doubt there is good reason to worry about machine AI and its power to destroy, but that power will come from human direction or bad programming, not from machine AI's consciousness.

These various considerations—language, art, NDEs, placebo effect, artificial intelligence—must suffice as the evidence that the human mind cannot be reduced to mere matter. Together they provide strong reasons to believe that the soul is spiritual and, by extension, could not have been created using Darwinian evolution or any other such process, but must in each case be created *ex nihilo* by God.

The Will

In Saint Augustine's psychological analogy of the Trinity, the Son is reflected in our ability to think. Our ability to will—to choose freely—is then seen as a reflection of the procession of the Holy Spirit from the Father and the Son.[54] Our human will is closely related to our understanding, because it can be thought of as deliberately choosing, that is to say, choosing after thinking about the various options before us. It is a mysterious faculty, because it allows us to love God who is love, but also to fall away from Him. And that is exactly what happened in the rebellion of Adam and Eve, which we call original sin.

Let us begin by taking a look at what Saint Thomas had to say about the knowledge of Adam in the garden of Eden:

> In the natural order, perfection comes before imperfection, as act precedes potentiality; for whatever is in potentiality is made actual only by something actual. And since God created things not only

for their own existence, but also that they might be the principles of other things; so creatures were produced in their perfect state to be the principles as regards others. Now man can be the principle of another man, not only by generation of the body, but also by instruction and government. Hence, as the first man was produced in his perfect state, as regards his body, for the work of generation, so also was his soul established in a perfect state to instruct and govern others.

Now no one can instruct others unless he has knowledge, and so the first man was established by God in such a manner as to have knowledge of all those things for which man has a natural aptitude. And such are whatever are virtually contained in the first self-evident principles, that is, whatever truths man is naturally able to know. Moreover, in order to direct his own life and that of others, man needs to know not only those things which can be naturally known, but also things surpassing natural knowledge; because the life of man is directed to a supernatural end: just as it is necessary for us to know the truths of faith in order to direct our own lives. Wherefore the first man was endowed with such a knowledge of these supernatural truths as was necessary for the direction of human life in that state. But those things which cannot be known by merely human effort, and which are not necessary for the direction of human life, were not known by the first man; such as the thoughts of men, future contingent events, and some individual facts, as for instance the number of pebbles in a stream; and the like.[55]

Saint Thomas's conception of the first man illustrates an important truth about original sin. Adam was truly guilty of sin, because God had given him the capacities and the freedom to resist the serpent's temptation. He wasn't a powerless reed blown about by the wind. But imagine, in Darwinian terms, an evolving primate lineage grunting its way toward self-consciousness—gradually, generation after generation—and then finally, with the birth of Adam, crossing some barely perceptible line. It would be an insult to justice to maintain that death and all sorts of other evils are the proper punishment for such a pathetic creature's misdeed, all the more so when we reflect that the process of natural selection would have primed man to be grasping, opportunistic, and self-seeking. But that is the Darwinian picture, not the biblical one. To safeguard God's justice, Adam needed to be fully aware of what he was doing, and fully equipped to meet and resist the temptation if he so chose. And according to Saint Thomas in the passage above, Adam was so equipped.

Some aspects of this passage are unlikely to resonate with the modern reader, even if he has understood the philosophical language. "Act" and "potentiality" usually need a course in Aristotelian metaphysics to be understood properly. But even once we grasp that the "act" and "potency" passage just means that you can't give what you don't have, we may still remain puzzled, perhaps with the same kind of unease that we experience when we read the words of Christ in the Gospel: "A disciple is not above his teacher, but every one when he is fully taught will be like his teacher" (Luke 6:40). If that is the case, how do we ever make progress in human knowledge? Surely Adam did not know calculus, nor did the rabbis in our Lord's time know quantum mechanics! The solution is in understanding that what Aquinas and Jesus speak of does not pertain to mathematics, the empirical sciences, and technology, but to a right understanding of the relations between God and man and among people.

Ultimately the doctrine of original sin remains a mystery that requires an act of faith. It is revealed to us in the third chapter of Genesis in terms of a serpent and the fruit of the tree of the knowledge of good and evil. The temptation rings true. Partake of the forbidden fruit, the serpent reassures them, and "you will be as gods, knowing good and evil." But with that knowledge comes sin and death. The Church does not oblige us to believe that there was actually a talking serpent—only that the devil was actively involved. But we are obliged to believe that Adam and Eve were our first parents, and they were subject to a moral test, which they failed, fully knowing what they were doing. Only in this way does the ensuing punishment make sense.

8. Anti-Theist Darwin and His Useful Instruments

> *Evolutionism is really more of a paradigm or methodology than a theory. For its present-day supporters, the important thing about evolution is that it was due to natural causes.*
> —Giuseppe Sermonti[1]

> *Whoever, therefore, is not enlightened by such splendor of created things is blind; whoever is not awakened by such outcries is deaf; whoever does not praise God because of all these effects is dumb; whoever does not discover the First Principle from such clear signs is a fool.*
> —St. Bonaventure, *The Soul's Journey into God*[2]

I WANT TO DO A COUPLE OF RELATED THINGS IN THIS CHAPTER. First, we will look at Darwin's beliefs closely enough to debunk the common notion that he was just an objective observer following the evidence in conceiving his grand theory. Rather, he was a materialist looking for a plausible means of excising God from natural history. Next, we will examine some prominent Darwinists and other evolutionists, particularly those who self-identify as Catholics. Whether they know it or not, they are Darwin's useful instruments because they aid and abet what atheistic philosopher Daniel Dennett described as a universal acid, one that "eats through just about every traditional concept and leaves in its wake a revolutionized world-view."[3] I especially want to look at what happens to belief in original sin if one accepts the Darwinian picture as true. My contention is that Darwinism is an acid—in this Dennett was right—that, if embraced, dissolves both clear reasoning and right belief.

Charles Darwin—Materialist

In a letter to Asa Gray, shortly after the publication of the *Origin*, Darwin says:

> I had no intention to write atheistically. But I own that I cannot see, as plainly as others do, & as I sh^d wish to do, evidence of design & beneficence on all sides of us. There seems to me too much misery in the world. I cannot persuade myself that a beneficent & omnipotent God would have designedly created the Ichneumonidæ with the express intention of their feeding within the living bodies of caterpillars, or that a cat should play with mice.[4]

The problem of evil is, no doubt, a real theological difficulty. For many it is the most persuasive argument against the existence and goodness of God. The Christian answer is the Cross of Christ—a difficult answer to a difficult question. Darwin had been studying for the Anglican ministry as a young man, so one might be tempted to pity his loss of faith in light of the cruel biological evidence. But if one looks more closely at his life, one will see that he was a materialist from a very early age. In fact, materialism ran in the family.

Charles's grandfather, Erasmus Darwin, had no use for any religion. Benjamin Wiker characterizes him as a Deist skeptic, political radical, and scientific adventurer, with a "polite Epicurean disrespect for traditional sexual morality."[5] His most celebrated work, *Zoonomia*, an early sketch of the theory of evolutionary transmutation, was an international success, with many American editions and translations into German, Italian, French, and Portuguese. Erasmus added three scallop shells with the motto *E Conchis Omnia* (all things from shells) to the family coat of arms as a sign of his belief in common descent. Although Erasmus presented himself as a Unitarian, Samuel Coleridge, after a visit with Erasmus, concluded that the man was an atheist.[6]

Darwin's father, Robert, continued to use the device *E Conchis Omnia*. Like Erasmus, he too seems to have been an atheist, although he was quiet about it and supported the Anglican Church. He was astute enough to have seen the connection between atheism and the terror of the French Revolution, a calamity which he was eager to keep out of England. So Charles received some formation in Unitarianism, the religion of the smart set in Britain: skepticism with a veneer of respectability. Robert at first sent Charles to study medicine in Edinburgh where he and his father Erasmus before him had also studied medicine, but it soon became clear that Charles was not cut out to be a physician. "So it was in all good British conscience," Wiker writes, "that Robert Darwin could propose, and Charles accept, the notion that a doubting Whig could matriculate in a Tory religious institution, and take up the life of a country parson. There he could continue with his 'shooting, dogs, and rat-catching,' and not be a 'disgrace' to himself and his family."[7]

Although Charles's days as a student of medicine were short, there is good evidence to suggest that they were important to his intellectual development as a materialist. Charles arrived in Edinburgh in the fall of 1825, with his brother Erasmus. When Erasmus went off to London to study in the fall of 1826, Charles was befriended by Robert Edward Grant, who was sixteen years older and already an accomplished biologist. Adrian Desmond and James Moore, in their biography of Darwin, write that in meeting Grant, "Darwin was coming under the wing of an uncompromising evolutionist.... Nothing was sacred for Grant. As a freethinker, he saw no spiritual power behind nature's throne. The origin and evolution of life were simply due to physical and chemical forces, all obeying natural laws."[8]

Grant was a member of the Plinian Society, a club for students of natural history, which had been founded in Edinburgh in 1823. Soon after Darwin met Grant, he petitioned for membership and was accepted into the Society, in November 1826. In an officially Christian country, the Society was a hotbed of heterodoxy. On the very evening that Darwin joined, he heard William Browne refuting Charles Bell's *Essays on the Anatomy of Expression*, in which Bell had argued that the human face was specifically endowed by the Creator to express emotion. Browne thought that this was anatomical chauvinism and that there were no essential differences between human and animal faces. The week after, Darwin heard a paper by William Greg, arguing that "the lower animals possess every faculty & propensity of the human mind." In March 1827, Browne returned to give a lecture in which he argued that the mind and consciousness were merely operations of the brain. Michael Flannery reports that "this was seen as so potentially dangerous that it was struck from the Society's minutes."[9] Darwin left Edinburgh shortly afterwards. He had attended eighteen of the possible nineteen meetings that academic year.[10]

All of this heady materialism was likely reinforced in Darwin's mind by his friendship with Grant. In his autobiography, Charles downplays the effect that this relationship had on his evolutionary thought, as well as the effect that his grandfather Erasmus's work had on him,[11] but there are reasons to doubt all this. As Wiker notes, Grant sought out Charles precisely because Charles was the grandson of the evolution-promoting Erasmus Darwin. It was clear that Grant was interested in evolution, and as notes in Darwin's copy of the *Zoonomia* indicate, Charles was more than a little interested in his grandfather's work himself.[12] Grant had Charles do some research on polyps, as part of his investigation into evolution. He then presented this work without crediting Charles, which may explain the break in their relationship. But his materialistic outlook remained with the young man, despite Charles's agreeing to study for the Anglican ministry.

Towards the end of his life, Darwin would usually decline meeting guests who sought him out in his estate at Down. Certainly he was careful not to say anything direct about his faith or metaphysical views. So it is strange that he should have welcomed the English atheist Edward Aveling and Europe's "fiercest materialist," Ludwig Büchner, when they called on him. During the visit, he questioned Aveling as to the wisdom of being so aggressive in pushing atheism on the populace. This was consistent with his letter to George Darwin, quoted at the beginning of the Introduction to this book. Better let materialist ideas corrode the faith of Christians slowly than wake Christians up and force them to make the "wrong" choice. During the same visit, Darwin said that he had not given up Christianity until he was forty years old.[13] This was a patent lie. His notebook C, written in 1838 when he was twenty-nine, has the following entry:

> Thought (or desires more properly) being hereditary it is difficult to imagine it anything but structure of brain hereditary., analogy points to this. – love of the deity effect of organization. oh you materialist! – Read Barclay on organization!!... Why is thought being a secretion of brain, more wonderful than gravity a property of matter? It is our arrogance, it [is] our admiration of ourselves.[14]

Indeed, even earlier, by the time he left Edinburgh, Darwin may already have been well on his way to materialism. The persistence of his arguments to trace various human traits—thought, religion—to animal instinct is just a continuation of the lectures he heard at the Plinian Society. Darwin's *The Expression of the Emotions in Man and Animals* (1872) was a justification of William Browne's talk of November 1826.

Flannery summarizes it well: "In the end, he [Darwin] would vindicate them all—Browne, Greg, and Grant—when his theory was finally constructed and published. Darwinian evolution, far from being a scientific theory, is 'one long argument' in favor of non-dogmatic atheism. He came to this not by his theory but by his fellow Plinians, who filled his young impressionable mind with ideas ill-suited to a sober weighing of objective facts."[15]

Wiker comes to the same conclusion about Darwin's uncompromising materialism: "Insofar as Darwinism has swallowed up evolution into itself, the evolutionary theory partakes of the deep antitheistic bias that Darwin built into it. It in fact does lead to atheism because it was designed to do so. The enormous push that secularization received from Darwinism should be proof enough that the theory of evolution so understood destroys belief in God."[16]

This short summary of the development of Darwin's materialism does not pretend to be a substitute for a biography of a complex human being. During the course of his studies for the Anglican ministry, Darwin was impressed with Paley's *Evidences*, no doubt trying to acclimatize himself to his father's choice of vocations for him. He felt compelled to tell his cousin Emma, before he married her, that he had religious doubts. Emma, a believer in God, thought that his honesty would at last lead him to faith and went on to marry him. Several years into their marriage, she worried that he was focusing so much on his work that he was closing himself off from coming to faith in God. Darwin was touched but did not take her advice. He kept this letter and appended a short note to it: "When I am dead, know that many times, I have kissed and cryed over this."[17]

Like any human being, Darwin was not totally self-consistent and could be moved at times to waver in his worldview. A year before his death, Darwin read the *Creed of Science*, by William Graham. The book examined new scientific ideas, including Darwinism, and tried to see how they could be reconciled to moral and religious ideas. Upon reading the book, Darwin wrote to Graham, "You have expressed my inward conviction, though far more vividly and clearly than I could have done, that the Universe is not the result of chance."[18]

When a Christian finds himself facing challenges to his faith, he is encouraged to say, "I believe; help my unbelief!" (Mark 9:24). Similarly, it has been noticed that there are very few atheists in foxholes. Perhaps Darwin's comment to Graham betrays some sympathy for a deistic conception of the divine. But such an idea of God is very far from the Christian notion of God, who is aware of every sparrow and lily of the fields. At times Darwin seemed to waver between a thorough-going deism on the one hand, far removed from Christian theism, and on the other hand, pure materialism, a view still further removed from Christian theism. What is unquestionable is that his theory has functioned to marginalize God from natural history and to serve as a support for materialism and atheism.

Pierre Teilhard de Chardin, S. J. (1881–1955)

Teilhard has been dubbed "the Catholic Darwin," although, strictly speaking, he was not a Darwinist. He was a paleontologist with his own particular understanding of evolution. As a young Jesuit studying theology in England, Teilhard was already enamored of natural history and evolution when he read Henri Bergson's *L'Évolution créatrice*.

"The only effect that brilliant book had upon me," wrote Teilhard, "was to provide fuel at just the right moment, and very briefly, for a fire that was already consuming my heart and mind."[19] This understates the book's impact. It proved influential on Teilhard's outlook on matter, life, and energy.

Bergson was the philosopher whose lectures at the Collège de France were responsible for giving enough hope to Jacques and Raïssa Maritain not to go through with their mutual suicide pact. The two students were in the grip of despair engendered in them by everything else they were hearing in the lecture halls of French academia. Fortunately for the Maritains, they could accept the lifeline that was Bergson's lectures, and then move on to Catholic truth interpreted through Thomistic philosophy. Teilhard, on the other hand, picked up Bergson's penchant for spawning neologisms—noosphere, hominization, etc.—as a means of appearing to say something profound. His writings caught the imagination of many religiously minded people who were looking for meaning in a world which, according to the prevailing view in the academy, was an eternally meaningless interplay of chance and necessity. Teilhard's thought can perhaps be summarized as a vision of the universe self-organizing into higher and higher levels of complexity until it reached its final destination—the Omega point—which he endowed with some connection to Christ.

No doubt, Teilhard had—and continues to have—Catholic admirers. The most positive Catholic assessment I have encountered comes from the pen of Msgr. Bruno de Solages, Rector of the Institut Catholique de Toulouse, who praised the "magnificent coherence" of Teilhard's view of evolution: "not materialistic, but essentially spiritualistic, not pantheistic but theistic, not deterministic but directed by God, not immanent but requiring the transcendental, not anti-Christian, but leading logically to the Christian supernatural." He continues:

> Within the framework of a Universe no longer cyclical as was the Aristotelian system—a great clock which eternally moves—but the Universe of modern science—which is an evolutionary Universe in progress, and one which, unless it be radically absurd, must necessarily go in a certain direction—Father Teilhard de Chardin

successively demonstrates the personal immortality of souls, and the existence of a personal God, the motivating force of all this evolution.... And we find that the heart of these two demonstrations is really nothing else basically but the Aristotelian and Thomistic principle, *desiderium naturae non potest esse inane*. "The desire of nature cannot exist in vain." Only, in place of applying this principle merely to the desire of a spirit, considered in its individuality, it is now extended to the totality of spirits, which is considered in the evolutionary perspective of the whole of modern science, as an actual term of the Universe.[20]

On the other hand, Fr. Raymond J. Nogar, O. P., had this to say about Teilhard's posthumously published *The Phenomenon of Man*: "Professional philosophers call it bad philosophy, professional theologians call it bad theology, professional poets call it bad poetry, professional scientists call it bad (mystical, which is worse) science, and, whatever its rhetorical advantage, professional dialecticians call it impossible dialogue."[21]

Outside of the Catholic world, some critics could be even more scathing than Fr. Nogar. Peter Medawar, a hard-core evolutionist, wrote a review of the *Phenomenon of Man*:

> It is a book widely held to be of the utmost profundity and significance; it created something like a sensation upon its publication a few years ago in France, and some reviewers hereabouts have called it the Book of the Year—one a Book of the Century. Yet the greater part of it, I shall show, is nonsense tricked out by a variety of tedious metaphysical conceits, and its author can only be excused of dishonesty only on the grounds that before deceiving others he has taken great pains to deceive himself. *The Phenomenon of Man* cannot be read without a feeling of suffocation, a gasping and flailing around for sense. There is an argument in it, to be sure—a feeble argument abominably expressed—and this I shall expound in due course; but consider first the style, because it is the style that creates the illusion of content, and which is in some part the cause as well as merely the symptom of Teilhard's apocalyptic seizures.[22]

Alasdair MacIntyre was also not impressed when the book was published in English in 1959:

> *The Phenomenon of Man* is a disappointing work... This book illustrates the banality of the too general. Father Teilhard de Chardin stuns the reader with assorted information, speculates on its significance, and draws his wide ranging conclusions without ever giving the reader a feeling that anything has been genuinely grasped. There is a particular opiate effect in the reading of all those books which attempt to summarize millennia. The content of Father Teilhard de Chardin's work is remarkably similar to the thesis which Sir Julian [Huxley] has himself advanced so often... Paradoxically what is missing from the picture altogether is history, and man is above all a historical being... It remains true that the attempt to turn the concept of evolution into a metaphysical key to the universe is one of the graveyards of the intellect, and the present work is merely one more testimony to that fact.[23]

In order to keep the discussion grounded, a few historical details about Teilhard de Chardin are in order. After he was ordained in 1911, he took up studies in paleontology and geology in Paris but did some fieldwork in England in the company of Charles Dawson and Arthur Smith Woodward. In 1912, Dawson and Woodward announced they had found fossilized bones of what they claimed was a missing link between man and apes, near Piltdown in East Sussex. The creature was given the scientific name *Eoanthropus dawsoni*, but it is usually referred to as Piltdown man. Not everyone, however, was convinced of the reconstruction or significance of the fossil, so more work was necessary.

In August 1913, Teilhard joined Dawson and Woodward in scouring through the spoil heaps of the gravel bed where the skull had been found, in an effort to find the canine teeth that had been missing from the jaw. It was Teilhard who found a candidate tooth, which, according to Woodward, fit the jaw perfectly. This was considered enough evidence to place the exhibit in the British Museum, where it remained until 1953. It was finally removed when it became clear that the fossil was a combination of a modern human skull and the jaw of an orangutan, cleverly put together by a forger who knew exactly what paleontologists expected from a missing link. By the time that it had been exposed as a fraud, it had been the subject of over 500 doctoral dissertations.[24] In 2016, an extensive scientific review pointed to Charles Dawson as the likely perpetrator.[25] Stephen Jay Gould, on the other hand, thought that Teilhard and Woodward were responsible.[26]

Soon after finding the tooth, Teilhard returned to France, where he served with distinction in the First World War. He received his doctorate in 1922 and started lecturing at the Institut Catholique in Paris. But he soon ran into trouble with his Jesuit superiors because of a lecture he gave in Belgium in 1922, in which he questioned the very possibility of Adam and the Garden of Eden and proposed a redefinition of original sin. His superior ordered him to sign a declaration disavowing these propositions and to correct the lecture, which the superior then sent to Rome along with Teilhard's signed retraction. While waiting for a response from Rome, Teilhard kept developing his controversial theses on original sin. As a result, Fr. Włodzimierz Ledóchowski, the superior general of the Jesuits, ordered him to sign a further declaration disavowing the controversial theories; and he also forbade him from giving any more lectures at the Institut Catholique.

At this point Teilhard went to China to look for more fossils of possible missing links. He was part of Davidson Black's research group, which found a molar which, in Black's estimation, had both apelike and human features. This supposed link came to be known as Peking Man, now Beijing Man, or *Sinanthropus pekinensis*. The bones of more individuals eventually showed up. But along with the bones, there was also evidence of technological innovations that seemed to be beyond the abilities of any animal other than man. In the end, despite Teilhard's arguments to the contrary, the consensus was that Beijing man was just an ancient member of the genus *Homo*. As Michael Chaberek puts it: "As we see, Teilhard's name appears in the scientific literature mainly on the occasion of the two most unreliable 'discoveries' regarding human origins ever. Piltdown man turned out to be a hoax, whereas Beijing Man was just a very ancient man whose skills and culture did not differ substantially from today's primitive peoples."[27]

Chaberek appreciates that there is something coherent and holistic about Teilhard's philosophical writings, which makes them attractive, but questions whether they have any connection to reality. It is clear that Church authorities thought that Teilhard's writings were dangerous. Bergson's *L'Évolution Créatrice* had been on the Index of Prohibited Books since 1914. In 1927, the Holy See denied the *Imprimatur* to Teilhard's *Le Milieu Divin*. In 1933, the Jesuits ordered him to resign from his post in Paris. In 1939, his *L'Energie Humaine* was added to the Index of Prohibited Books. In 1941, Teilhard asked for permission to publish *Le Phénomène Humain*. This request was denied. In 1947, Teilhard was banned from lecturing on philosophy. Not one to be discouraged, in 1948, Teilhard again asked for permission to publish *Le Phénomène Humain*. He was summoned to Rome by his Jesuit superiors, banned from publishing his work, and even denied office space at the Collège de France. In 1949, he was told not to publish *Le Groupe Zoologique*. And in 1955, the year he died, the Holy See forbade him from participating in the International Paleontological Congress.

Teilhard bequeathed the rights to his works to various publishing houses. This was against his vow of poverty, which according to canon law deems that whatever assets an individual might possess are the property of the religious order. Nevertheless, his will prevailed[28] and *Le Phénomène Humain* appeared in France in 1955. It was then published in English, with an introduction by the arch-Darwinist Julian Huxley, the grandson of Thomas Henry Huxley, just in time for the Darwin centenary in 1959.

The popularity of these published works led the Holy See to issue a *monitum* (warning) in 1962, through the Congregation of the Holy Office: "Prescinding from a judgment about those points that concern the positive sciences, it is sufficiently clear that the above mentioned works abound in such ambiguities and indeed even serious errors as to offend Catholic doctrine."[29]

Despite these warnings, Teilhard still has his supporters in the Church. In 1981, on the centenary of the birth of Teilhard, there was a celebration at the Institut Catholique in Paris, presided over by

Archbishop Paul Poupard. A largely laudatory letter, signed by Cardinal Casaroli on behalf of John Paul II, was read out.[30] This prompted many to think that the *monitum* no longer applied. So a clarification was published in the *Osservatore Romano*: "After having consulted the Cardinal Secretary of State and the Cardinal Prefect of the Sacred Congregation for the Doctrine of the Faith, which, by order of the Holy Father, had been duly consulted beforehand, about the letter in question, we are in a position to reply in the negative. Far from being a revision of the previous stands of the Holy See, Cardinal Casaroli's letter expresses reservation in various passages—and these reservations have been passed over in silence by certain newspapers—reservations which refer precisely to the judgment given in the Monitum of June 1962, even though this document is not explicitly mentioned."[31]

More recently, in 2009, Pope Benedict XVI made some positive comments during a homily in which he praised Teilhard's vision of the entire cosmos as a "living host."[32] It was not the first time that Joseph Ratzinger had said something positive about Teilhard. In his *Introduction to Christianity* (1969), Ratzinger turned to Teilhard as support for a contemporary explanation of Pauline Christology, which sees all the elect united in Christ, the new Adam, who will be all in all. "One can safely say," commented Ratzinger, "that here the tendency of Pauline Christology is essentially correctly grasped from the modern angle and rendered comprehensible again, even if the vocabulary employed is certainly rather too biological."[33]

Not everyone would agree that Teilhard sheds positive light on the mystery. C. S. Lewis dismissed Teilhard's writings as "pantheistic-biolatrous waffle" and "evolution run mad."[34] And Ratzinger too could be critical. As John Allen put it in the *National Catholic Reporter*: "In a commentary on the final session of the Second Vatican Council (1962–65), a young Ratzinger complained that *Gaudium et Spes*, the "Pastoral Constitution on the Church in the Modern World," played down the reality of sin because of an overly 'French' and specifically 'Teilhardian' influence."[35]

In 1948, Teilhard wrote an appendix to *The Phenomenon of Man* entitled "Some Remarks on the Place and Part of Evil in a World in

Evolution." In this short essay, Teilhard admits that he does not treat evil in his works: "True, evil has hitherto not been mentioned, at least explicitly. But on the other hand it inevitably seeps out of every nook and cranny, through every joint and sinew of the system in which I have taken my stand." So, where does it come from? "On this question, in all loyalty, I do not feel I am in a position to take a stand: in any case, would this be the place to do so?"[36]

This is a deeply disappointing admission coming from a Catholic priest. Original sin is an essential teaching of the Church. The Cross is the answer, albeit mysterious, to the great mystery of evil. How can a serious Christian possibly write about humanity without addressing the reality of evil? How can a Catholic priest feel that he is not in a position to identify where evil comes from?

It seems to me that the people most in favor of Teilhard are impressed by his positive vision, which, he insists, is derived from science. They look upon it as a great attempt to synthesize faith and science, because they tend not to understand science. Those who understand science are not so easily fooled, because most of the "scientific" element in Teilhard's work is verbiage masquerading as science. But even if the science were real, it would be a bad starting place for theology, because science keeps changing, whereas God's revelation does not.

Joseph Ratzinger and Henri de Lubac are two accomplished scholars and men of faith who saw something positive in Teilhard's efforts. We have already encountered some negative assessments, but it may be instructive to present two more.

First, we have the witness of Thomas J. J. Altizer, who was not a Catholic: "It is true that Teilhard occasionally and inconsistently introduces traditional Christian language into the pages of *The Phenomenon of Man*, but this fact scarcely obviates the truth that virtually the whole body of Christian belief either disappears or is transformed in Teilhard's evolutionary vision of the cosmos."[37]

Second, and from within the Catholic fold, there is Étienne Gilson. He exchanged several letters with Henri de Lubac, who, like Teilhard, was a Jesuit. Gilson clearly thought that Teilhard was a great danger to Catholic orthodoxy:

> You can't get any benefit or any enlightenment from thinking about Teilhard. The ravages he has wrought, that I have witnessed, are horrifying. I do everything I can to avoid having to talk about him. People… use him like a siege engine to undermine the Church from within (I'm not kidding); and I, for one, want no part in this destructive scheme.[38]

In the same letter, Gilson laments:

> What worries me is rather that while all our Christian theologians, starting with your allegorists, developed their theologies from meditation on the Scriptures, Teilhard, grounded in his evolutionist consciousness, built his theology from a meditation on science. Thus it amazes me when people stick his Pauline sayings about the "cosmic and evolving" Christ on the bulletin board, because if sin and grace are not the very foundation of the Epistle to the Romans, I don't know anything about Saint Paul. Myself, I'd a hundred times rather be a Lutheran than a Teilhardian. The real Saint Paul is the one whose Christ *tollit peccata mundi* [takes away the sins of the world], not the type that would cause a theologian to reject, as the vicious belief that everything has always been the same, the very existence and even the possibility of sin. Whoever does not believe in sin has no right to the Christ Saint Paul believed in.[39]

Gilson did not think that Teilhard was teaching a doctrine that could be censured, because it seemed to him that Teilhard's thought has not attained the minimum state of self-consistency for it to be classified as a doctrine. But he was clearly worried about Teilhard.[40]

In an earlier letter to de Lubac, Gilson recounts that he met Teilhard twice. The first time was when Teilhard was baptizing a baby for which Gilson stood as godfather. Gilson was impressed by Teilhard's exercise of his priestly ministry on that occasion.[41]

The second meeting took place at a conference near New York in 1954, a year before Teilhard's death:

> We hadn't had much to say to each other, except naturally, for the usual small talk. In our first encounter, though, Father dealt me a one-two punch that left me reeling for three days afterwards.

I never regained my poise. As soon as he spotted me, he strode straight up, his pleasant handsome face lit by a smile, and clapping his hands on my shoulders, he said to me point blank: "Hey, can you tell me if anybody's ever going to give us a scoop on this religionless [metachristrianity is the exact neologism] Christianity we've all been waiting to hear about?" Taken aback, I spluttered a bit, finally managing to make an awkward joke of which I think the sense was that I already had enough to do to get a grip on plain Christianity, and so we left it at that.[42]

Later in the day, Gilson passed by Teilhard and saw that he was engrossed in his breviary.[43] Teilhard was clearly an enigma—a dangerous enigma.

Theodosius Grygorovych Dobzhansky (1900–1975)

When one reads works that try to sell evolutionary thought to Christians, one is almost sure to come across references to Theodosius Dobzhansky. Dobzhansky was born in Ukraine and emigrated to the United States in 1927, already as an accomplished entomologist, studying fruit flies. He became a US citizen and taught and did research in various American Universities: Columbia, Caltech, and UC Davis (as emeritus).[44]

In 1973, towards the end of his life, he published an essay entitled "Nothing in Biology Makes Sense Except in the Light of Evolution."[45] The essay was a criticism of anti-evolution creationism and a call to espouse theistic evolution. In arguing that life was united and diversified, Dobzhansky noted that the protein cytochrome C was found to be similar in a wide range of life-forms, from yeast to fish to man. Dobzhansky also pointed to the various species of fruit flies in the Hawaiian Islands as examples of evolution.

There is nothing that intelligent design cannot handle in these assertions. The fruit-fly diversification is clearly a microevolutionary scenario. And a closer look at the cytochrome C sequence rather shows very separate branches of life. Why, for example, should the human, fish, reptile, and bird cytochrome C sequences all be about 65 percent different from the cytochrome C sequence in bacteria?

And why should the yeasts be more or less equidistant from every other more complex life-form? Michael Denton used this evidence to argue that there were separate archetypes in biology that could not have evolved from one another in a continuous manner.[46] Denton goes into considerable depth on this score. The point here is simply that Dobzhansky's argument is far from conclusive.

Dobzhansky saw in Teilhard a kindred spirit, as one who sees evolution as the way by which God brings the cosmos to perfection, so much so that he became the president of the Teilhard Association in 1969.[47] Dobzhansky's reference to the "light of evolution," which he uses to end his essay, probably borrows from a passage in *The Phenomenon of Man*: Evolution is "a general condition to which all theories, all hypotheses, all systems must bow and which they must satisfy henceforward if they are to be thinkable and true. Evolution is a light which illuminates all facts, a curve that all lines must follow."[48]

Christopher Howell notes that "Dobzhansky's views on religion were idiosyncratic and highly personal, and the extent to which he held to specific Orthodox doctrines is unclear. Although he was open about his sympathy for religion and his interest in philosophical issues, he kept much to himself."[49] Jitse van der Meer, in his study of Dobzhansky's religious beliefs as they related to his understanding of evolution, says, "Privately, Dobzhansky described himself as a communicant member of the Eastern Orthodox Church though not in the full sense." This was based on a letter he wrote to his daughter and son-in-law in 1956.[50] This self-description makes it legitimate, at some level, to cite Dobzhansky as a poster child for Christians adopting Darwinism. For example, Ernst Mayr, an atheist, wrote, "Famous evolutionists such as Dobzhansky were firm believers in a personal God. He would work as a scientist all week and then on Sunday get down on his knees and pray to God."[51]

Yet Dobzhansky's evolutionary views of Christianity were not orthodox in any ordinary sense. Indeed, Francisco Ayala, who had been his student and was with Dobzhansky at the moment of his death, thought Dobzhansky did not believe in a personal God, contrary to Ernst Mayr's opinion.[52] We need not take Ayala's word as final. Whether or not Dobzhansky believed in a personal God is a separate issue from what all can agree on: Dobzhansky could be called an orthodox Christian only in the sense that he considered himself a member of the Orthodox Church.

George Coyne, S. J. (1933–2020)

Pope John Paul I, in his very brief pontificate, appointed Fr. George Coyne to head the Vatican observatory, which the Jesuits have staffed since the sixteenth century. He comes into the evolution story because of a dustup he got in with Cardinal Christoph Schönborn, whose July 2005 op-ed piece in the *New York Times*, "Finding Design in Nature," asserted that to accept Darwinism was to abdicate reason. In the piece, Schönborn argued that both John Paul's and Benedict's views regarding evolution and design had been misrepresented, and that there is clear evidence from their writings that both rejected the Darwinian vision of man arising from purely mindless natural processes.[53]

If the cardinal was upset by the misrepresentations of John Paul's and Benedict's views, Coyne was livid at the cardinal's intervention. In an essay, "God's Chance Creation," published in the English Catholic magazine *The Tablet* in early August 2005, he accused the cardinal of muddying the already "murky waters of the rapport between the Church and science" by attacking the "best of modern science."[54] The article, however, is a clear example of muddled thinking on Coyne's part.

"Science," he declares at one point, "is completely neutral with respect to philosophical or theological implications that may be drawn from its conclusions." Yet he goes on to say that science has a bearing on our understanding of divine omnipotence and omniscience. So is it neutral or not? One wonders whether he is even aware of the contradiction as he trumpets his Teilhardian vision of the cosmos.

I have discussed this particular exchange at much greater length in an article published by *Touchstone* in June 2006.[55] Here, I just want to highlight the logical incoherence that Coyne falls into. He desperately wants to cling to Darwinism, but to do that, he has to keep philosophy and Catholic theology out of the picture.

Kenneth Miller

Kenneth Miller describes himself as a Catholic and is recognized by many Catholic institutions. The Society of Catholic Scientists awarded him the Saint Albert Award in 2017 in recognition for his "contributions to science education and public understanding of science." The short biography also notes that on earlier occasions he had received the Gregor Mendel Medal from Villanova University and the Laetare Medal from Notre Dame University. A website focused on spotlighting Catholics scientists writes of him, "He has been outspoken in explaining and defending the theory of evolution and its consistency with the Catholic faith."[56]

As we saw in an earlier chapter, he has been less than forthright in his blog debates with Lehigh University biologist and design proponent Michael Behe. Miller's ideas of co-option to explain the evolution of irreducibly complex biological machines such as the bacterial flagellum are not up to challenging intelligent design in any serious way. So here, rather than revisit the technical debate, I will focus on Miller's claim to be in line with Catholic teaching.

I am not impugning his sincerity. But in order to be Catholic, one has to believe the basic teachings of the Faith. Yet reading his various works on ultimate questions in the light of evolution, one gets a sense that Miller cannot possibly hold orthodox Catholic positions on some fundamental teachings, such as divine omniscience and (possibly) the immateriality of the soul. Admittedly, one cannot find outright denials of these teachings in his works, but one certainly gets the sense that something is amiss.

Take, for example, a statement found in multiple editions of Miller's textbook *Biology*: "Evolution works without either plan or purpose... Evolution is random and undirected."[57] He is widely quoted on the internet as saying, "I see a planet bursting with evolutionary possibilities, a continuing creation in which the Divine Providence is manifest in every living thing. I see a science that tells us that there is indeed a design to life, and the name of that design is evolution."[58]

If evolution is the "design" of Divine Providence, then evolution must be directed, or providence has no meaning. So we encounter a contradiction. Miller wants it one way for his Darwinist textbook sales and another for his more Christian audience.

The word "providence" comes from the Latin *pro*, meaning "before," and *videre*, meaning "to see." A provident person sees what is necessary beforehand and provides it. So if Divine Providence were really at work, then God, at least, should know that the evolutionary process, which he started, would end in a particular way. What then do we make of Miller's approval of Stephen Jay Gould's contrary view of the evolutionary process? Miller writes: "No question about it. Rewind that tape [of evolution], let it run again, and events might come out differently at every turn. Surely this means that mankind's appearance on this planet was not preordained, that we are here not as the products of an inevitable procession of evolutionary success, but as an afterthought, a minor detail, a happenstance in a history that might just as well have left us out. I agree."[59]

What are we to think? Did God, according to Miller, know from all eternity that man would arise? It seems not. So on this view, God is not omniscient, and we human beings are not directly willed by Him. This may not be problematic for deists or pantheists, but it does not square with Catholic orthodoxy.

One gets the same kind of uneasy feeling from another passage from *Finding Darwin's God*: "And so to ask the big question, do we have to assume that from the beginning he [God] planned intelligence and consciousness to develop in a bunch of nearly hairless bipedal African primates? If another group of animals had evolved to self-awareness, if another creature had shown itself worthy of a soul, can we really say for certain that God would have been less than pleased with His new Eve and Adam? I don't think so."[60]

I have already remarked that this kind of thinking is difficult to reconcile with divine omniscience and divine providence. What is new in this passage is that Miller sees a purely unguided material process giving rise to self-awareness. If self-awareness can be achieved for a purely material entity, what is one to think of the soul that Miller mentions? It seems that it would be superfluous, except to give Miller some Christian credentials by not denying its existence altogether.

It is interesting to read reviews of Miller's *Finding Darwin's God*. Miller is impressed by the anthropic coincidences in physics. He has no patience for those who try to explain them away by invoking an infinity of universes. This led Barry Palevitz, an atheist critic of his work, to write, "Wait! Is Ken Miller, irreducible complexity's worst nightmare, using exactly the same arguments as Behe, except that instead of designing biochemical pathways, Miller's deity plays dice with quarks?" Miller points out that physics is not biology. But Behe, in his review of Miller, makes it clear that Miller is really very muddled. Miller, for example, concedes that "the indeterminate nature of quantum events would allow a clever and subtle God to influence events in ways that are profound but scientifically undetectable to us. Those events could include the appearance of mutations, the activation of individual neurons in the brain, and even the survival of individual cells and organisms affected by the chance processes of radioactive decay." But this is exactly in line with ID theory—not orthodox Darwinism. Behe concludes: "'Evolution is not rigged' and 'God exercises the degree of control he chooses' are compatible only if God chooses to exercise no control. To be consistent, Miller either has to give up Darwinism to allow for the active God of Christianity or settle for a deistic God at best to allow for Darwinism."[61]

Theistic Evolution—Christopher Baglow

Teilhard, Dobzhansky, Coyne, and Miller are representatives of what has come to be known as theistic evolution. They have tried to put a divine spin on what was designed by Darwin to be a metaphysics that excluded God. Richard Dawkins gets it. Richard Lewontin gets it. Christians tend to be slow.

To be sure, there are many different meanings of theistic evolution. But unless one is willing to acknowledge that God is guiding the process with a goal in mind and that the empirical evidence makes it possible to detect the work of an intellect, in other words, that ID is the best explanation of the phenomena, one is going to end up with much confusion and usually a lack of concern for Christian orthodoxy.

I am heartily in agreement with philosopher J. P. Moreland's assessment of this common type of theistic evolution: "[It] is intellectual pacifism that lulls people to sleep while the barbarians are at the gates. In my experience, theistic evolutionists are usually trying to create a safe truce with science so that Christians can be left alone to practice their privatized religion while retaining the respect of the dominant intellectual culture."[62]

The Templeton Foundation is very keen on theistic evolution.[63] It supports the institution BioLogos, founded by Francis Collins, the one-time head of the human genome project. The organization's mission is to reconcile evolutionary biology and the Christian faith. The Templeton Foundation also gave a grant for the publication of *Faith, Science, and Reason: Theology on the Cutting Edge*, by Christopher Baglow. At first sight, the book appeared to be the perfect text for the philosophy of science class I teach to seminarians. But I will not be using it because the section on evolution is pure Darwinian propaganda.

Right at the start comes the denial of the crucial distinction between microevolution and macroevolution: "Some distinguish macroevolution, the emergence of new species, from microevolution, the minor incremental changes that occur within a species. Ultimately, as Darwin realized, these are the same kinds of changes, and so it is preferable to speak of speciation."[64] Yes, preferable if you do not want students noticing that there is abundant direct evidence for microevolution and little if any direct evidence for macroevolution. You definitely want to get rid of that distinction, even though it is one that Darwinists in good standing have employed for generations.

To be sure, Baglow tells us that Darwin did not have all the answers, but he quickly assures us that, thanks to Mendelian genetics, the theory is now complete. Moreover, "This 'neo-Darwinian synthesis' is the basis of the modern understanding of the origin of species and is accepted by virtually all biologists today."[65]

In a subsection on "Darwin's Faith, Darwin's Doubt," Baglow summarizes: "While many religious people accepted Darwin's theory of evolution, there were many who did not. Indeed, there are many Christians today who bitterly oppose Darwinian evolution. As we have seen, some of them think that it is contrary to the book of Genesis; others have theological or philosophical objections to it." Conveniently unmentioned are those of us who oppose Darwin's theory on scientific grounds. The scientific problems are enormous, and they will not go away.

Baglow does not inspire confidence in his comprehension of the relevant science. He writes, "Genes are the blueprints for the construction of living bodies through the assembly of protein molecules, 'the workhorse of life's processes.'"[66] The truth of the matter is that no one has a clue as to what makes a mouse a mouse or a horse a horse. Yes, DNA codes for proteins and has many other yet-to-be-discovered functions. But the blueprint idea is a metaphor of the past, now debunked, as discussed in Chapter 2.

Baglow also tells us that genes and fossils are converging in producing "a huge amount of interlocking and mutually supportive evidence that confirms that all living things on earth, including human beings, are indeed related to each other in a giant 'web' of life."[67] It is difficult to know exactly what he means other than that most human beings have an intuition that the great variety of plants and animals on the earth are recognized as being alive. If he means to say that classification based on fossils and classification based on molecular biology are producing a single, coherent tree of life, then he is not aware of the accumulating evidence against this holy grail of Darwinism. (Once again, see Chapter 2.)

Baglow's section "Evolution: Theory or Fact?" concludes that it is "very solidly established." To this end, he turns to Theodosius Dobzhansky's paper, "Nothing in Biology Makes Sense Except in the Light of Evolution." And this provides him with the perfect opportunity to say that a Jesuit priest and scientist, Pierre Teilhard de Chardin, was the inspiration for such a grand vision.

Baglow then paints Darwin as coming from a devout family that said its daily prayers. Darwin, he suggests, was a sincere student for Anglican orders who was disappointed when it became clear to him that Paley's argument for God's existence lost its force in the face of natural selection. Darwin's faith, Baglow says, was shaken because he made the mistake of thinking that God was the "how" of nature rather than the "why." Yet neither of Baglow's claims is convincing. Baglow's how/why distinction would seem to leave God nothing to do in creation, even at the point of the Big Bang, rendering him even less active and involved than the God of modern deism. And Baglow seems oblivious of Darwin's skeptical family background, his early immersion in materialism, and his obsession with keeping God out of the picture even in the creation of man.

Contrary to what Baglow tells us in his discussion of the "The Catholic Church, the Theory of Evolution, and the Human Soul," the Church has been wary of the theory of evolution since its inception. He apparently has not read Michael Chaberek's *Catholicism and Evolution: A History from Darwin to Pope Francis* (2015), which helpfully puts into one volume all the relevant church documents, and which was available well before Baglow's second edition (2019). The Church, aware of the bad press of the Galileo affair, chose not to make any great public pronouncements, apart from the Council of German Bishops meeting in Cologne in 1860, which condemned Darwin's theory. But behind the scenes and quietly, the Church, through various organs, silenced the enthusiasm of some of her children for evolutionary theory: Raffaello Caverni, Dalmace Leroy, John A. Zahm, Henri Dorlodot—all priests who had argued for what today is called theistic evolution.[68] None of these is well-known, so Baglow could be forgiven for not discussing their cases. But any Catholic attempting to argue for the truth of theistic evolution should know that Teilhard de Chardin was on several occasions silenced by his Jesuit superiors and by the Magisterium of the Church.

Baglow is too faithful a Catholic to ignore the teaching on original sin and the proscription against polygenism, both in the older sense, according to which various races derive each from distinct first parents (instead of all from Adam and Eve), or in the sense the term is sometimes employed in theological discussions, namely the modern idea that the entire present human population evolved from a population of hominins rather than from a single first couple. About original sin, Baglow rightly says that we do not know the details of what happened, that Genesis reveals the truth in language that is not to be taken at a naively literal level. But polygenism is a problem. He opines that Pius XII, in his condemnation of it, might have been overly concerned in the immediate aftermath of the Second World War about the negative effects of racism, which might be encouraged by the idea that different human races arose from different parents; better to insist on our common humanity through Adam than to open up the door to racism. Baglow also points out that the language of *Humani generis* indicates that the condemnation of polygenism might be provisional, for the Pope said: "Now it is in no way apparent how such an opinion can be reconciled with that which the sources of revealed truth and the documents of the Teaching Authority of the Church propose with regard to original sin, which proceeds from a sin actually committed by an individual Adam and which, through generation, is passed on to all and is in everyone as his own."[69]

In this last surmise, Baglow might be correct. If original sin could be accounted for in some other way, then perhaps Catholics could embrace it. The problem for Baglow and other theistic evolutionists is that there is just too much in the biblical texts and Church tradition to make it likely that such a solution can be found. Chaberek cites a speech given by Pope Saint Paul VI in 1966 to a symposium of theologians: "It is evident that you will not consider as reconcilable with the authentic Catholic doctrine those explanations of original sin, given by some modern authors, that start from the presupposition of polygenism, which is not proved."[70] As Chaberek notes, this intervention did not stop people from trying; and Baglow clearly has hopes in this direction.

More recently, Baglow has published *Creation: A Catholic Guide to God and the Universe* (2021), a volume in the Engaging Catholicism Series. Once again he tries to steer away from monogenism, noting that an encyclical is not necessarily an infallible teaching. He quotes Kenneth W. Kemp to argue that *Humani generis* was toned down to make wiggle-room on this question possible: "Archival materials on the drafting of the encyclical were opened to researchers on March 2, 2020, and I was able to review these documents in the week before the imposition of a general curfew made continued work impossible.

The archival records make clear that the non-definitive language was deliberately chosen over the stronger language of early drafts of the encyclical."[71]

Baglow even provides a way to understand what traditional theology would list among the effects of original sin. He quotes Kenneth Miller to explain why the brain is responsible for so much confusion: "The circuitry of the brain is in fact a poorly integrated mixture of the truly ancient, the very old, and the relatively new all working side by side."[72] Baglow then draws his own conclusion: "In summary, despite being one of the great wonders that evolution produced, our biological and neurological inheritance has its shortcomings. This explains a great deal about why humans act irrationally and contrary to goodness, why racism and other forms of bias seem to crop up throughout history, why male human beings find it so hard to treat women as equals and not as objects, and why wars and the killing of the innocent are constants in human history. There is nothing from the scientific perspective that should make us surprised by this."[73]

Baglow tries to be faithful to the Church's teachings, so he attempts to reconcile the "scientific" view with the Tradition. He begins by saying that the various traits that we share with the animals, such as desire for survival, health, and pleasure, are not intrinsically evil, but pre-moral. That is where evolution got us. Then we were offered the preternatural gifts, including integrity and immortality. This was God's plan A for bringing us to share in His glory. But our ancestors somehow lost this gift. The removal of these preternatural gifts puts us back to the state where evolution brought us to. That, of course, would be a cause for despair, were it not for plan B—God's sending His own Son to take on our human nature and offer us the gift of eternal glory through His death and resurrection.

Baglow continues in the same vein in his newer book. There we read: "Using the examples of blind ideologies and racism, we will explore how so many of the worst human tendencies have been illuminated by modern cognitive science, which shows that the animal nature we inherit from evolution is at the root of our tendencies toward both moral goodness and moral evil."[74]

To his credit, Baglow takes original sin seriously. Admittedly it is a very difficult topic and it is easy to go astray. Saint Augustine devoted much time and energy to thinking about it; and while there is much to gain from those reflections, certain aspects of his thought have come to be rejected by the Church, such as the eternal damnation of infants who die prior to baptism. Some of the more obvious difficulties of original sin are (a) that it is called sin—a personal act—but it is imputed to persons who did not commit it and (b) its mode of transmission is rather mysterious—the Church has defined that it is through generation rather than imitation, but does that mean that God creates imperfect souls?

Furthermore, although there is not an overabundance of biblical texts that demand a single Adam, the father of the human race, to be solely responsible for the catastrophe, there are some important passages of scripture that cannot be overlooked. There is, of course, Genesis 3, although in the next chapter, Cain is marked so as not to be killed by others, implying that the world was already populated, presumably by human beings. The traditional answer to that objection is that Adam and Eve had many other children, as the book of Genesis itself mentions, albeit in a later chapter. There is also Saint Paul's first letter to the Corinthians: "For as in Adam all die, so also in Christ shall all be made alive" (1 Cor. 15:22). In the Acts of the Apostles, Saint Paul tells the philosophers on Mars Hill: "And he made from one every nation of men to live on all the face of the earth" (Acts 17:26). The commentators usually see this as a reference to the unity of the human race, arising from Adam. But there are some textual variations to which one can appeal to evade the common reading. One of these texts, for example, omits "from one" altogether.

Finally there is the passage in the letter to the Romans: "Wherefore as by one man sin entered into this world, and by sin death; and so death passed upon all men, in whom all have sinned" (Romans 5:12: Douai-Rheims Bible). One would think that this is clear, but there is a great dispute about the exact translation and its meaning in the whole context of the passage and its bearing on original sin. Just about all modern translations of the verse reject the Vulgate "in whom" (*in quo*), which Catholic theology emphasized until recent times, and which is why I had to resort to the Douai-Rheims translation in this case. So if one relies only on modern biblical exegesis, the scriptural case for monogenism is not as strong as one might have expected, given the detail in Genesis that Cain was afraid of an apparently significant number of people having it in for him, despite his being in only the second generation of humans following the first couple, and given the Genesis story of the Nephilim, discussed earlier.[75]

Nevertheless, the *Catechism* continues to call original sin "an essential truth of the faith"—and with good reason—in fact, two good reasons. First, "The doctrine of original sin is, so to speak, the 'reverse side' of the Good News that Jesus is the Saviour of all men, that all need salvation and that salvation is offered to all through Christ. The Church, which has the mind of Christ, knows very well that we cannot tamper with the revelation of original sin without undermining the mystery of Christ."[76] Second, "Ignorance of the fact that man has a wounded nature inclined to evil gives rise to serious errors in the areas of education, politics, social action, and morals."[77]

In addition, the *Catechism* tells us how to read the account of Adam's rebellion: "The account of the fall in Genesis 3 uses figurative language, but affirms a primeval event, a deed that took place *at the beginning of the history of man* [emphasis in original]. Revelation gives us the certainty of faith that the whole of human history is marked by the original fault freely committed by our first parents."[78] Some such rebellion is necessary to make sense out of God's being all-good, all-powerful, and the creator of man in the face of the evils in and around us, with the one certainty of our precarious life here being that it will end with the death of the body.

Just about all the difficulties associated with coming to understand original sin have been around for centuries. Now, only because the scientific consensus has bought into the grand Darwinian story, we are being told that we must rethink monogenism and original sin. We are also told that we must rethink our ideas about God's omniscience and omnipotence. George Coyne has said so explicitly. And it seems to me that Kenneth Miller believes the same thing, as does Christopher

Baglow. If we are really warped because our brains came out a mess through the evolutionary process, then what does that say about God's guiding the process, or starting one that was going to produce a flawed product that would have to be held together by a whole panoply of preternatural and supernatural Band-Aids?

Thomistic Evolution

Within the broader category of theistic evolution is a more specialized approach called Thomistic Evolution, led by "a team of Dominican friar scholars (Fr. Nicanor Pier Giorgio Austriaco, O.P., Fr. James Brent, O.P., Br. Thomas Davenport, O.P., and Fr. John Baptist Ku, O.P.) committed to the preaching of the Gospel, and, through the Thomistic intellectual tradition, providing answers to questions about faith and reason, and religion and science." So reads the back cover of their book *Thomistic Evolution: A Catholic Approach to Understanding Evolution in the Light of Faith*.

These scholars accept the current Darwinist picture, reject intelligent design, and try to interpret evolution in terms of Thomistic categories of instrumental causality. I will not dwell long on their efforts, because another Dominican, Michael Chaberek, has criticized their work in his *Aquinas and Evolution*. I will limit myself to one very telling observation.

In the first edition of *Thomistic Evolution* (2016), Austriaco argued that the human race's emergence from a population of hominins, rather than from a first couple, is a theological possibility for faithful Catholics. Science, he said, has progressed since Pius XII insisted that we all descend from one original set of parents. Moreover, in 2004, the International Theological Commission, presided over by then Cardinal Joseph Ratzinger, stated, "Catholic theology affirms that the emergence of the first members of the human species (*whether as individuals or in populations*) represents an event that is not susceptible of a purely natural explanation and which can appropriately be attributed to divine intervention." According to Austriaco, "This suggests that both" scenarios "remain viable theological opinions for Catholic theologians seeking to be faithful to the doctrinal tradition."[79]

A papal encyclical is usually accorded more authority than the speculation of theologians, but Austriaco needed some such permission to accommodate what to him was established science, namely, that "one would need to posit the existence of 10,000 original humans to properly account for the genetic diversity that we see among the seven billion human beings living today."[80] He then went on to speculate how original sin could be understood under this scenario.

There is no need, however, to go into the details of this scenario, because Austriaco had jettisoned it by the time of the next edition in 2019 and adopted monogenism. Why the about-face? Let us read Austriaco's explanation:

> This critical point in evolutionary history occurred about 100,000 years ago in Africa among a group of anatomically modern human beings when an individual hominin was conceived with the inherent neurocognitive capacity for language.
>
> How exactly this occurred will always be a matter for speculation. As Noam Chomsky has proposed, a single mutation acquired at conception could have altered the structure of an individual anatomical modern human's brain in a way that gave him the capacity for language. In my view, for reasons I cannot develop at this time, this capacity for language would have gone hand in hand with the capacity for abstract thought. As the first speaking primate, this individual would have also been the first rational animal. He was the first anatomically modern human to have the capacity to form abstract concepts, to reason, and therefore, to construct an internal map of his world. As he matured, he would have used this linguistic ability to speak to himself and to God. This first speaking human is the original human we call Adam.
>
> Moreover, since every human being today possesses the same linguistic capacity, each one of us must have inherited that capacity from him. Each one of us must therefore be descended from this first speaking human. Adam would not only have been the first speaking human, he is also the father of all speaking humans, which is all of us.[81]

We have several problems here. The first problem is the idea that a single mutation in DNA could be credited with giving us the ability to reason and speak. This is an absurdly tall order for a single genetic mutation, given what we know about the limits of such mutations. After all, the capacity for language and abstract thought is, qualitatively, one of the biggest leaps forward in the history of life—arguably as big as the origin of the first life and the origin of consciousness in the first animals. And we're to believe that a single mutation to a strand of DNA produced it? It flies in the face of common sense. It also presents a fitness problem. Before this magical mutation, the human form would be roughly as weak as it is now, but without our intelligence and capacity for speech. And as I have argued earlier, such relatively weak and unintelligent animals would not survive long in a dog-eat-dog world.

There is also a theological problem. What does Austriaco's new scenario do to the need for a spiritual soul? In his just-so story, the purported DNA replaces the spiritual soul as the explanation for our unique mental and spiritual capacities.

Next, philosophers of language usually take Ludwig Wittgenstein's argument against a private language seriously.[82] Why would this lucky primate ever think to develop a language, especially since no one else around would be able to speak? Next, what about his mute, pre-rational mate? What happens to all the rich theology of the body that draws its inspiration from Eve's being made from the side of Adam? If Adam had to wait for a child to come along before he could experience a truly human relationship, one would hardly blame him for thinking that existence was a burden and for turning his back on God.

It is depressing to see that a Catholic theologian who specializes in the interaction of faith and science should be so easily swayed by the latest speculations from science as to keep rewriting the very important theology of our origin and fall. And to be even more critical, Chomsky's surmise had been around for quite some time and was showing its age. Why pick it up in 2019, five years after the 2014 paper Chomsky co-authored, which stated that no one has any clue about the evolution of language?

The stakes are high. To recast original sin in some Darwinian form would require the Church to change some teachings that have long been considered *de fide*: doctrines which the Church presents as directly revealed by God and essential to the Catholic Faith. To change a doctrine of such authority is to signal a lack of belief in God's revelation and the Church's infallibility as a teacher of faith and morals. Man's changeable opinion, rather than God's revelation, would become the basis of the Church. Such a church would soon fade into oblivion.

A much better strategy, which I will describe in the final chapter, is to reject the broad claims of Darwinism on scientific, philosophical, and theological grounds.

9. Converging and Convincing Arguments

O felix culpa, quae talem ac tantum meruit habere Redemptorem!
Oh happy fault, which merited such and so great a Redeemer!
—Exultet

These pages have spanned many disciplines—biology, archaeology, geology, physics, cosmology, the history and philosophy of science, and Catholic theology. So in this final chapter, I will summarize the salient points, in an order and in a way that I hope lends itself to a clearer understanding of the arguments. This overview also will allow us to highlight some revealing connections and, here and there, extend the argument for design.

Darwinism was from the start designed to remove the direct action of God from the story of natural history, even if Darwin threw God an occasional sop, as when he inserted a suggestion in the second edition of *The Origin of Species* that the Creator had perhaps breathed life into one or a few forms, thereby kick-starting the evolutionary process (a sop he later privately denounced and said he regretted inserting into the book).[1]

Darwin, following in his father's and grandfather's footsteps, was becoming an anti-theist by his late teens and seems to have been settled into that view soon thereafter. The picture that emerges from a careful study of his biography is not that of someone pursuing the scientific evidence where it led and grudgingly embracing the theory of evolution despite the challenge it posed to his religious faith, but of a young man devising a theory to give a scientific basis for his materialistic view of the world.

Darwin would countenance no limitations on his theory. When Alfred Russel Wallace, who had independently come up with the idea of evolution through natural selection, wrote an article in which he argued that natural selection could not explain the higher human faculties and that there was an Overruling Intelligence guiding the process of evolution, Darwin left traces of his anger in the margin: "No!!! I hope that you have not murdered too completely your own & my child."[2] Thus we see that the essence of the theory, as Darwin conceived it, was the eclipse of God from natural history, and perhaps even his wholesale abolition.

Nevertheless, a scientific theory meant to justify atheism could still be true. That is why I have spent quite some time looking at the evidence for Darwinism. As we saw, the theory falls apart under scrutiny. Darwin was aware that the fossil evidence was against him. His theory predicted that fossils would show a slow, almost imperceptible, change over time. The fossil record of his day contradicted his theory. But he could live in the hope that as more fossils were discovered, his theory would be vindicated. Unfortunately for him, the fossil evidence has become more rather than less problematic for Darwinism, as paleontologists have gained more and more knowledge of life's history on earth from new fossil discoveries and improved techniques for understanding and dating those fossils. The Cambrian explosion of new life-forms was only one of several unexpected diversification events. It is now clear that soft-bodied animals and plants leave fossils, cutting off another explanatory escape route for Darwinists seeking to paint the abrupt appearance of new forms as an artifact of an incomplete fossil record. The world has been scoured for the fossils that Darwin believed must exist, but they have not been found.

Of course Darwinists, committed to their scientific paradigm, cling to their theory in the face of these developments. And their creativity in conjuring up creative patches exceeds even the Medieval astronomers with their appeal to ever more epicycles to account for the motion of the heavenly bodies within a geocentric model. That the Darwinists have countless such patches at hand is not to be doubted. That these patches do more than cover the truth should be duly considered.

Since Darwin's time, his theory has been upgraded to what is known as the neo-Darwinian synthesis by the addition of population genetics. Neo-Darwinism was trumpeted as the undoubted truth about life in time for the Darwin centenary in 1959. But it soon attracted the attention of top mathematicians who pointed out that the chances of assembling long specific biochemical molecules through an unguided process were infinitesimally small, even granting the power of natural selection to guide random genetic mutations. As the amount of information coded into DNA became known, and as the technology for manipulating genes and proteins made it possible to get a handle on the probabilities of blindly arriving at functional proteins, the case against Darwinism has grown stronger. Douglas Axe, for example, was able to show experimentally that, for one portion of a typical enzyme, the ratio of amino acid sequences that can fold (and so become biologically useful) to the total number of possible amino acid sequences for that given stretch is approximately one in 10^{77}. Given that the total number of organisms that have ever lived on the earth is of the order of 10^{40}, there is virtually no possibility that blind evolution could hit upon the myriad working proteins found in today's life-forms,[3] much less orchestrate them into the functional wholes characterized by interdependent systems of systems of systems, such as we find in plants and animals.

Strictly speaking, abiogenesis, the origin of life from inorganic chemicals, is not part of Darwin's theory of evolution by natural selection, nor its updated form, neo-Darwinism. Abiogenesis is, rather, its presupposition. Nevertheless, since it was my intention to argue against materialism and not just Darwinism, I spent some time looking at how research in this area has produced a long series of dead ends.

If some of the considerations in that section were too technical, common sense should come to the rescue. The simplest self-reproducing biological entity is necessarily more sophisticated than any human-made nano-technology. Even relatively simple organisms reveal an internal unity and functional integration beyond any mere manmade machine. Human engineers are, despite great effort, nowhere close to inventing a self-replicating machine of any size—that is, a machine that can build copies of itself that can build copies of itself, on and on.[4] Thus, while it did surprise many scientists how enormously complex even the simplest single-celled organisms are (viewed in Darwin's time as something like protoplasmic goo), it really shouldn't have, given the design sophistication required to achieve self-replication.

Eugene Koonin, no friend of intelligent design, has suggested that the odds of life arising by chance in our universe are less than one in $10^{1,018}$ (the latter a number starting with 1 and followed by 1,018 zeroes!).[5] But he says he has a way to overcome these odds: an infinity of universes. These other universes are, by definition, inaccessible to empirical verification, but he needs them as an escape hatch from the design implications of the mind-bendingly small odds of the chance origin of life. It should be clear from this tactic that the science-museum exhibits with warm ponds and lightning to suggest how the first cell arose by a natural process are just propaganda for materialism.

A further argument against Darwinism is based on the tactics of Darwinists. They insist that there is "an overwhelming consensus" in their favor and then aggressively sideline and silence dissenting voices, even when the dissenting voices are well-credentialed and highly placed scientists offering purely evidence-based critiques of modern Darwinism. Why employ such tyrannical techniques if Darwinism is supported by a real examination of the facts? And why are some specialists in these fields skeptical of Darwinism in the first place?

In arguing against Darwinism, I am not arguing against evolution in the general and non-controversial sense of substantial change over time in the history of life. There is unequivocal evidence for that. Some of this change can even be explained by natural selection. But much more of it cannot; and even materialists are producing some alternatives to the standard neo-Darwinian story: horizontal gene transfer, hybridization, evo-devo, and epigenetics are often mentioned.[6] But these naturalistic alternatives also face acute problems, evidenced by the fact that the various camps in this so-called "third way" movement (neither intelligent design nor neo-Darwinism but something else) shoot down each other's positions with cogent objections.

Opponents of intelligent design often accuse leading design theorists of being oblivious of these innovative third-way variations on evolutionary theory and of merely beating the dead horse of neo-Darwinism. And yet such leading design theorists as Michael Behe and Stephen Meyer, among others, have offered detailed critiques of these third-way strategies. If materialists have a powerful case against intelligent design, why misrepresent it in this and so many other ways? Creating a strawman of your opponent's argument is a reliable sign that one hasn't found a good way to take down the opponent's actual position.[7]

All those on the side of truth should be grateful that the influential philosopher Thomas Nagel, who confesses that he actively does not want there to be a God, came to the defense of the ID community's cogent critique of Darwinian materialism.

No doubt, the atheists are determined to find a naturalistic explanation for all of life and will insist that every complex life-form arose from some more primitive life-form. But that is going beyond the evidence. The more one looks into the gaps between higher-order classifications, the more one becomes skeptical of a single tree of life. Indeed the more molecular and morphological data that is available to construct "the tree of life," the holy grail of Darwinism, the more contradictory the exercise becomes. Instead of a tree, a tangled grove of bushes would be a better metaphor.[8]

One facet of biology that everyone seems to agree on is that the biological molecules are great storehouses of information. The dispute is about how the information got there. The materialists invoke the interplay of chance and a blind necessity. But the odds against that are too great. That is the negative argument against Darwinian materialism. But there is also a positive argument *for* intelligent design: In our uniform experience as a species, the only type of cause known to produce novel functional information is the creativity of mind—that is, intelligent design. Thus, the best explanation for the information found in nature is intelligent design.

Everyone agrees that life-forms appear to be designed. The proponents of intelligent design argue that this appearance is not misleading. Instead of looking to the chemical and physical properties of paper and ink to explain the composition of Hamlet, they look to an author. It is a much better way to understand the play. The same reasoning applies to the book of nature.

Orthodox academia makes life hell for anyone who comes out in favor of ID. Nevertheless, some very knowledgeable scientists have decided to pay the price of losing their prestigious positions in order to speak up about what the evidence is telling them—that life is designed. Most of them do believe in God, so it is easy to paint them as Bible-thumping "creationists." But that is misleading for multiple reasons. Some who left Darwinism for intelligent design did so as non-religious scientists, such as internationally renowned paleo-entomologist Günter Bechly, who only became a Christian (and in his case, Catholic) some time later.[9] Also, when the term "creationist" is thrown around to characterize proponents of intelligent design, many automatically think of persons who espouse a young earth and ground much of the argument in a particular way of reading the early chapters of the book of Genesis. That picture misses the mark. The contemporary arguments for intelligent design are grounded in public evidence revealed by science, with no appeal to special revelation; and most of the leading proponents of intelligent design accept that the universe is nearly fourteen billion years old.

The various scientists and philosophers associated with the ID community all see design in nature as something real and detectable. However, there are differences of opinion as to whether the designer is God and, if so, whether this God is the God of the Christians. There are also differences of opinion about whether universal common descent is true. Michael Behe, for example, thinks it likely is. Michael Chaberek and Stephen Meyer think that it is not.

Additionally, there are differences among proponents of intelligent design as to how the designer guides or intervenes in the process of change. There is little discussion of this in the literature, mainly because the focus of design theory is on determining whether something was designed or not, not on determining the methods employed by the designer. Michael Denton is looking for some laws of forms, imbedded by the designer in the fabric of nature from the beginning. This view is more structuralist in orientation and rejects Darwinian functionalism. And while his critique of Darwinism is cogent, it is difficult to see what these hypothesized laws of form might be.

Behe, as we saw, is open to the idea of a series of directed interventions, incremental and of the quantum mechanical type—all but undetectable except for its resulting in new species and a degree of novelty in far fewer generations than any purely material process could achieve. To illustrate the principle, consider a roulette wheel. We can be sure that some outside directing must be going on if a perfectly balanced roulette wheel were to produce the beginnings of the Fibonacci sequence time and again. But if God were subtly doing the directing, perhaps by subtly moving the hand that throws the ball, our searching for a specific mechanism for the bias in the roulette wheel would never reveal it. In such a case, God would influence the operation of the roulette wheel, contributing the information that the physics of the roulette wheel could never produce. Behe suggests that something like this may have been at work in the evolutionary history of life on earth.

Other ID theorists find this approach less appealing. Some of them prefer a more straightforward reading of God's creation of "kinds" and his creation of man from dust, and see little reason either from theology or paleontology to conclude that God worked subtly and incrementally.

Saint Thomas's distinction between God's direct actions in nature and His use of secondary causes is instructive in this context. God is the first cause of all, holding all creation in being. If He chooses to create a new species from scratch, then He would be acting as first cause. If, on the other hand, God chooses to use the natural unfolding of the laws of physics and chemistry to produce all the heavy elements in the universe via stellar nucleosynthesis, that would be an example of God working through secondary causation. Theistic evolutionists hold that God used the Neo-Darwinian process of random mutations and natural selection to produce life's diversity—a form of secondary causation. This book has argued that the evidence weighs heavily against such a view—evidence gleaned from scientific observation as well as from philosophical and theological reflection.

Those arguments do not negate something like Behe's evolutionary proposal, a directed form of evolution with God intimately involved in the creative process along the way, and in a way that clearly reveals that the outcome is not the result of a blind process. That being said, I do not want to speculate further as to how the Divine Intelligence might operate in relation to evolution. Suffice it to say, whether He filled the earth with life's diversity in an extremely gradual, incremental fashion or did it in great bursts of novelty remains a point of research and debate among design theorists, as does the question of precisely where God employed secondary causes versus direct causation. In any case, if God is involved in evolution, somehow He knows every detail and guides every step, directly or indirectly. Contra some theistic evolutionists, He is not surprised by any particular outcome.

I am always disappointed to hear Catholics disparaging ID. Perhaps they are afraid of giving credence to a God-of-the-gaps type of argument that will be rendered false by further progress in science. It is wise to be wary, but one must be careful not to throw the baby out with the bathwater. The ID position is a powerful means of undercutting materialism—a chief modern obstacle to faith—and to helping prove the existence of God. The Catholic Faith, according to the First Vatican Council, obliges us to believe that human reason can demonstrate the existence of God, although the decree does not tell us how to do it. The *Catechism*, however, clarifies that reason can come to know God through "'converging and convincing arguments,' which allow us to attain certainty about the truth."[10]

Moreover, if we look to the Bible, we often find various created things and their harmony singing the praises of God. "The heavens are telling the glory of God; and the firmament proclaims his handiwork,"[11] the Psalmist tells us. ID takes the same kind of perspective. It includes the flowers of the field and the birds of the air and adds DNA and RNA, proteins, optimally designed biological structures, and other features our ancestors in the Faith would not have known about.

The examples come from biology, because our ancestors recognized the marvel that is life, and here I want to emphasize that ID is in historical continuity with Scripture. But the reader also should remember from a discussion in Chapter 3 that one can detect other designs in nature. I mentioned, for example, the cosmological constant, the balance between the four fundamental forces, and the suitability of various elements for processes from the biological to the planetary to the astronomical. These observations were not something that the early Christians knew anything about. I can also add, as Jay Richards and Guillermo Gonzalez point out in *The Privileged Planet*, that we find ourselves optimally placed not only to be alive but also to make a range of scientific discoveries, and the very suite of features optimized for habitability are also optimized for scientific discovery. If that doesn't look like a set-up, it's hard to imagine what would.

The intelligent design argument starts one down a path toward knowledge of God's existence, a path in line with the biblical and patristic approach to establishing the existence of God. It follows a hermeneutic of continuity. It is also related to Saint Thomas's fifth way, from teleology. This is significant, because most moderns find the first four of Thomas's ways either incomprehensible or wrong or uninspiring. Kant dismissed them altogether, whereas he had great respect for the argument from teleology.

If we reject ID, this not only causes a rupture with tradition, but also makes it difficult to know where to start on the road to God when engaging with many modern secular people. Many are convinced that science has disproven the existence of God and miracles. Therefore, they regard the Bible as a collection of fairy tales masquerading as history, and the Tradition as so much benighted Medieval hocus-pocus. Yet I think that even some Catholics enamored of Darwinism sense the problem with jettisoning ID altogether. Kenneth Miller, for example, is impressed with the fine-tuning of the universe—the anthropic coincidences—evident in the laws of physics. I too am impressed with this. But I simply add this to the other "converging and convincing" arguments without throwing out the ones from biology.

We also examined the claims that ID could not possibly be true because of (1) purported imperfections in various organs, and (2) the presence of death, disease, and evil in nature. To the first I replied that the commonly cited "bad designs," such as the backward wiring of the vertebrate eye, are in fact clever design compromises involving unavoidable engineering tradeoffs. To demand otherwise is to forbid God from creating anything.

As to the existence of death, disease, and evil, there are two things to consider. First, even an evil design still needs a designer. The death chambers at Auschwitz were designed by evil human beings. So even if one concludes definitively that some design in nature is truly evil, one is still saddled with the recognition that it was designed. Second, even if some things are indeed examples of true natural evil, we should be very careful about assigning blame. We do not have the full picture. God does. What we do know is that we live in a fallen world. This is a mystery, which we might want to deny, but it is real. We are engaged in a spiritual battle that spilled over into the material world through the rebellion of Adam and Eve—a spiritual struggle, moreover, that predated their rebellion, due to the rebellion of Satan and his angels. We are now capable of designing very dangerous things such as weapons of mass destruction, perhaps even biological weapons, which sometimes get used. Perhaps the fallen angels had a hand in perverting God's creation. We do not know all the details, but sinful rebellion left a huge rift in creation that does not invalidate the ID argument. Just as there is enough goodness in the world for us to recognize it as the production of God, so there is enough design in life to recognize the action of a designer.

More directly against the God-of-the-gaps objection, I pointed out that the universe is permeated by gaps. It must be so by God's design. The Catholic Church teaches that each human soul is immediately created by God. So here is one gap that will never be filled by physics and chemistry. In addition, whenever a human being chooses to do something, the laws of physics must be accommodating enough to make it possible for him to move his arm or not. Further, there are angels who do things in the material world. And there are miracles. A materialist ideologue such as Richard Lewontin might want to keep the divine foot out of the door, but God has designed the world precisely to make it possible for Him to interact with the world and to have beings made in His image shape the world through their actions. The "gap" is not a lacuna in knowledge which will be filled up. It is an intrinsic feature of the cosmos. Thus, the science focused exclusively on particles and forces cannot possibly account for every aspect of reality.

If the only damage Darwinism had done was eliminate a sense of wonder and block the path to the Creator that begins with a recognition of design in nature, it would already have done major damage to our humanity. But it does still more damage through its false story of the origin of man. So I turned to examining the evidence for our originating from a slow, mindless process from some primate ancestor.

Clearly there is insufficient fossil evidence for a gradual evolutionary transition between an ape-like creature and the earliest member of the genus *Homo* with our basic anatomical structure, *Homo erectus*. The supposed ancestral fossils are usually just skulls (often partial) and jaw bones. And often what exists is badly crushed, giving "restoration" efforts room for the pro-evolution prejudices of the paleontologists to create transitionals in the workshop. Although we clearly resemble apes in many ways, there are huge structural differences in the attachment of the head, the hip joint, the feet, and the shape of the chest, not to mention in the unimaginably complex organization of our brains. These changes are not going to happen by tweaking a few genes. Far more is required.

There is evidence that the genus *Homo* has been around for two million or more years. The DNA evidence for the mating of Neanderthals with Denisovans and with modern humans is quite plausible, and the proposed dates for these trysts go back to more than 50 Kya. I tend to agree with Piotr Lenartowicz that these ancient lineages are "ecotypes" within the genus *Homo*. They could all be descendants of Adam and Eve, which many tens of thousands of years of microevolution shaped in different ways. Whereas in today's world there is only one ecotype of man—*Homo sapiens*—there is no proof that *Homo heidelbergensis*, *Homo neanderthalensis*, *Homo erectus*, etc., were not fully human.

The one problem with this interpretation is that evidence of clearly human culture does not stretch back to the time of the earliest *Homo erectus* fossils. Most or all of a set of key markers of early human culture—complex tools, clothing, jewelry, burial customs, musical instruments, the use of fire, highly skilled representative art—arrive hundreds of thousands of years later in the geological record, and in some cases, over a million years later. If *Homo erectus* was fully human, like all its descendent forms (e.g., *Homo heidelbergensis*), descended from Adam and Eve and thus made in the image of God, then this is a puzzle.

I suggested a few considerations without claiming to have solved the problem: low population densities, long periods of cultural degradation and stagnation, and loss of archaeological evidence by the flooding of coastal plains—the inhabited areas—by water from the receding glaciers at the end of the various ice ages since the dawn of man. There is also the further problem that the older artifacts would naturally be more degraded than more recently produced evidence. There is, for example, little reason to think that *Homo erectus*, who was capable of producing Acheulean stone tool technology, could not also produce at least primitive wooden tools, and yet there are none in the fossil record before 780 Kya—very possibly because it is extraordinarily unlikely for any wooden tool to survive even a hundred thousand years, much less many hundreds of thousands of years. How many artifacts of human culture from these distant times have simply dissolved to dust?

I also looked at the evidence for human exceptionalism and against a materialistic account of the mind and brain. That humans are exceptional among the animals is evident to anyone with eyes. The Judeo-Christian theological explanation for this widely observed fact is that humans are unique among the animals in being created in the image of God, with an immaterial soul. As for countering the materialistic account of the mind/brain, we looked at many lines of evidence pointing to the reality that the mind is not just an aspect of the brain, and that thoughts are not chemical secretions or networks of firing neurons.

Other evidence we considered included human language, near-death experiences, the neuroplasticity of the brain, the placebo effect, terminal lucidity, the limits of general artificial intelligence, and the implications of Wilder Penfield's findings during open brain surgery on conscious subjects. And on the theological side of the ledger, we contrasted the evolutionary view of man as grunting his way to language and rationality with the traditional view of Adam as someone who had to know exactly what he was doing if his banishment from the Garden of Eden was not going to be an affront to divine justice.

I am grateful to Stephen Jay Gould for his candid admission that the critics of Darwinism sometimes have many good points. But he wasn't above smearing critics of Darwinian theory with accusations of bad faith. "Creationism," he warned, "is a mere stalking horse or subsidiary issue in a political program that would ban abortion, erase the political and social gains of women by reducing the vital concept of family to an outmoded paternalism, and reinstitute all the jingoism and distrust of learning that prepares a nation for demagoguery."[12] He wrote that in 1983. Today, the same accusation would be directed against the ID community, with the "stalking horse" now also said to have the goal of overturning legal and cultural victories pertaining to LGBTQ+ issues. This is not surprising, because Darwinism is the creation myth of modernity. With no mind to oversee evolution, there is no higher lawgiver and hence no higher law; we can define ourselves as we please. The secular creation myth is taught in science museums and publicly funded education, so it is not surprising that by now it has deep roots.

It is disappointing to watch Catholics fall for this deception. Teilhard had precursors among Catholic priests trying to accommodate evolution, but they were largely silenced in discreet ways. Teilhard was more persistent and willing to labor under various prohibitions. He also had a way with words that proved irresistible to many people—a grand vision of the universe evolving towards its final consummation in the omega point. The crucial theological issue that he did not bother to address was original sin. The crucial scientific issue he ignored was the growing body of evidence against modern evolutionary theory.

Today, several influential Catholics think that they have received a carte-blanche to proselytize for Darwinism from Saint John Paul II's allocution to the Pontifical Academy of Sciences. These include George Coyne, Kenneth Miller, Christopher Baglow, and Nicanor Austriaco. It is true that John Paul II said that evolution was more than a theory, that is to say, that there was much evidence for the transformation of life-forms over time. But then he specified: "And to tell the truth, rather than speaking about the theory of evolution, it is more accurate to speak of the theories of evolution. The use of the plural is required here in part because of the diversity of explanations regarding the mechanism of evolution, and in part because of the diversity of philosophies involved. There are materialist and reductionist theories, as well as spiritualist theories."

So directed evolution under Divine Providence might be true, but where does that leave Darwin? "As a result, the theories of evolution which, because of the philosophies which inspire them, regard the spirit either as emerging from the forces of living matter, or as a simple epiphenomenon of that matter, are incompatible with the truth about man." Clearly Darwinism is out of bounds. Darwin would not have been surprised by the condemnation. The question is, why do the aforementioned Catholic evolutionists not get it?

Two points are especially disappointing in reading the Catholic enthusiasts for modern evolutionary theory. The first is that they buy into the notion that matter can be rearranged so as to give rise to self-consciousness. This is a huge unwarranted assumption. Even atheist philosophers of mind, such as Thomas Nagel, see the problem. Catholic Darwinists, on the other hand, talk about God's waiting to see what would come out of the evolutionary process and then, upon seeing a creature attain self-awareness, plopping a soul into it. One wonders what was left for the soul to do.

The other big problem confronting the Catholic Darwinist is the Church's doctrine of original sin. It is an essential teaching of the Church. No doubt it is a difficult teaching with many facets that need to be understood in a coherent way—biblical revelation, God's justice, human biology, and centuries of observation of human nature. Some of these teachings have been defined by an ecumenical council—Trent. They are considered *de fide*—revealed by God and to be accepted as such by Catholics. If there is any wiggle-room, it is razor-thin. And it does not seem to me that there is any wiggle-room when it comes to monogenism.

If these Catholics are proposing overturning these teachings, they are in fact saying that the Church is wrong on essential *de fide* dogmas. They have made science a superior standard to God's revelation. They have set themselves above the teaching of the Church, which has been established by Christ with a promise that the gates of hell would not prevail against it. If they really think that the Church is wrong on such central questions, they should leave. To demand that the Church change its dogma is to ask the Church to self-destruct.

As Fr. Piotr Lenartowicz put it, "In spite of the often repeated declarations to the contrary, there is no way to conciliate the contemporary scientific and the fundamentally religious 'Weltanschauung.'"[13] The good news for the Catholic is that the contemporary scientistic worldview is false. If the world at large has not heard of Darwinism's demise, it is because the work of the ID community has been dismissed and ridiculed, and a collection of fawning Catholics and other Christians with prominent platforms have rushed to the defense of Darwinism and tried to change the Church's teachings to accommodate it. Yet in the midst of this sad state of affairs, the faithful need not worry that the Church has been wrong on the origin of life, man, and sin. On this, the old truths remain unshaken. We can lament that our first parents, Adam and Eve, rebelled against God. But we can sing with the Church: "*O felix culpa, quae talem ac tantum meruit habere Redemptorem!* Oh happy fault which merited such and so great a Redeemer!"

Appendix A.
Some Notes on Transformism

When Darwinists find themselves facing strong arguments against their theory, their longstanding tactic is to brand their opponents as believers in special creation—witness the confrontation between mathematicians and Darwinists at the Wistar Institute in 1966, discussed in Chapter 2. It is an effective tactic, because in many people's mind, *special creation* carries with it the idea of absolute fixity of species as we might find them in a bird guide today. On this understanding, proponents of special creation can be caricatured as believing that, in the beginning, God made the first pair of blue jays and also the first pair of Canada jays and Stellar jays. The implication is that anyone who wants to distance himself from fixity of species had better stop criticizing Darwinism because it is the only alternative, so never mind the lack of empirical evidence for it.

But one would be hard-pressed to find any special creationist holding such a narrow understanding of fixity of species. Even young earth creationists allow that when the Bible says, "God made the beasts of the earth according to their kinds," the word "kinds" designates a category much broader than the Linnean level "species." "Kind" is what is sometimes called "natural species" and corresponds roughly to the Linnean classification levels of genus or family. In this way, one can account for the undisputed facts of Darwin's theory of evolution—varieties of dogs, cats, and other domestic animals and plants, Galápagos finches, cichlids, bears, bacteria, and many more—without insisting that these all need a special creative action on God's part. The limit of genus or family corresponds nicely with the limits of evolution that Michael Behe has argued for on purely empirical grounds in *Darwin Devolves*.

So does this mean that we have to invoke special creation to go beyond the level of these natural kinds, to explain the origin of the separate orders, phyla, and kingdoms in the Linnean scheme? Here we run into disagreements within the anti-Darwinian or pro-ID side. Michael Chaberek thinks that one should be as faithful to the biblical text as possible and say that, yes, God needs to create the natural species or kinds *de novo*, that is to say, not from preexisting living beings. He thinks that Thomistic/Aristotelian metaphysics demands this. He objects to universal common descent and to my use of the word "intervention" when I speak of how God could have guided an evolutionary process that could account for universal common descent.[1] He thinks that the new forms appearing in time are special acts of creation and not just intervention in the sense of God's providential guiding of creation. Michael Behe, on the other hand, thinks that God could have intervened by adding the necessary information into the molecules of life through some quite gradual and all but undetectable quantum processes.

I tend to lean towards Michael Chaberek's point of view on universal common descent, but since the Church has not ruled definitely on this point and science has not provided a definitive answer, I am content to remain agnostic as to whether all life-forms alive today share a common ancestor. However, I do make an exception. I am all but fully convinced that, when Genesis says, "The Lord God formed man of dust from the ground," the word "dust" refers to some inanimate material rather than an evolutionary line of primates. My preference is based on the theological presupposition that the Bible is first and foremost a revelation of the relationship between God and man, so the origin of Adam and the subsequent origin of Eve as it is presented in Genesis should provide details for theological reflection that we dismiss at our peril. There are various reasons to interpret the "days" of creation to mean something other than 24-hour periods, but there is no reason to jettison the special creation of man from "dust."

The empirical evidence for God's involvement—his special creative actions in the generation of new kinds—is found in the various explosions of new phyla, subphyla, and classes in the fossil record, diverse creatures that cannot be linked in any convincing way to what came beforehand. And there is the further observation that although humans have been able to do some amazing things with selective breeding of various animals, no one has been able to transform, say, a dog into a cat.

If nature indeed lacks the potential to effect these "beyond kinds" transformations, does that rule out universal common descent? Not quite. If, for example, God took fresh creative action so that salamanders hatched from fertilized fish eggs—the "hopeful monster" scenario of Richard Goldschmidt—or even, say, He engineered the transition in a dozen generations, and he did this sort of thing throughout the history of life, then we might end up with a fossil record that looks like universal common descent. So if *any* sort of parent-to-offspring continuity counts as descent, then universal common descent could still be true. But if one insists that universal common descent is tied to *naturally occurring* reproduction, then universal common descent appears untenable. Too many considerations in paleontology, molecular biology, bioengineering, probability theory, on and on, mitigate against it.

The near universality of the genetic code, an oft-cited argument for common descent, cannot resolve the issue. God could use the same genetic code over and over again, just as electrical engineers use the same basic circuits to construct the most diverse systems, and a software programmer may use the same programming language to write many different kinds of software. Additionally, the genetic code, as it turns out, isn't truly universal. Variations in the genetic code have begun to turn up that are hard to explain in materialistic terms.[2]

Another argument that comes up from time to time is that God could have front-loaded evolution into the initial conditions and physical laws and constants at the Big Bang. Like a master billiards player, God could have set up the table so that at all the balls would be sunk with one shot—seeming like magic but really all following from the initial setup. This is an interesting idea that should not be scoffed at and dismissed out of hand, but there is precious little evidence for that scenario, and good evidence-based reasons to doubt it. Most fundamentally, we have no evidence that the known regularities of physics and chemistry can produce novel information, and novel information is required, in reams, to produce even the simplest biological life.

Why am I introducing these various scenarios of how present species might have arisen from past species? It's not to try to present an exhaustive list of possibilities or to exhaustively adjudicate among them. Rather, it is to show that accepting transformism need not mean buying into the Darwinian paradigm of a purportedly blind process that somehow produces an information-rich biosphere.

I am aware that some neo-Thomists are wary of terms such as "interventions," without which it is difficult to speak of transformation, because using "intervention" seems to suggest that God had to monkey with His creation, which is something disrespectful to His original creation. But, as Logan Gage points out, "Thomas himself, far from being worried about intervention, thought that there was good reason to think that God purposefully 'intervenes' in nature, writing that 'the divine power can at times work apart from the order assigned by God to nature, without prejudice to His providence. In fact, He does this sometimes to manifest his power. For by no other means can it be made manifest that all nature is subject to the divine will, than by the fact that sometimes He works independently of the natural order: since this shows that the order of things proceeded from Him, not of natural necessity, but of His free will.'" Gage says that he prefers Thomas's language "working independently of the natural order" to "intervention," yet it is an intervention, nevertheless. He concludes that "those Thomists who decry 'interventionism' may not be as Thomistic as they think."[3]

Thus, despite the objection of some neo-Thomists, I maintain the term "intervention," and I cannot find more neutral terms than "transformism" and "intervention." These terms, taken together, allow the possibility that yesterday's life-forms may have given rise to what we see today, allowing it without specifying exactly how that occurred.

Appendix B.
When Did Adam and Eve Live?

I am grateful to critical readers of my manuscript who were skeptical about my openness to a very early Adam and Eve, and in particular my willingness to regard Adam and Eve as being created as much as 2 million years ago, with *Homo erectus* (but not *Homo habilis or Homo naledi*) and later *Homo* species among their descendants. Do I really want to insist on that? Am I not sticking out my neck for something that might turn out to be false? Am I not perhaps too quick to dismiss the significance of the great leap forward evinced by the cultural and technological artifacts that date from much later, say some fifty thousand years ago? Am I unnecessarily introducing novelties into Catholic theology?

I cannot of course claim to have the univocal backing of science, because, as I have stressed, the narrative for hominin origins in contemporary paleontology is a big mess. There's no consensus. I have merely tried to make some sort of sense out of the generally accepted physical data, guided by Catholic theology. There is a great anatomical gulf between *Homo erectus* and the ape-like primates that preceded them. The anatomy of *Homo erectus* is strikingly similar to ours, even if there are some distinctive differences. But they are the first unambiguously upright walkers of roughly our size, with a lifestyle not dramatically different from that of the hunter-gather *Homo sapiens* of the Mesolithic, and with a brain volume that, while a bit smaller than the modern, is within the range of variability for *Homo sapiens*.

And while we do not have artifacts demonstrating a flowering of human culture from them, there is evidence that they learned to control fire (if not start it), and quickly developed the capacity to work stones into tools using foresight and planning far beyond that of our closest living primates. They quickly spread throughout Africa, and Eurasia below a certain cold line, and may even have had the ability to build primitive boats. We also have good reason to believe that even if *Homo erectus* were creating wood and bone tools and weapons, it's unlikely such artifacts could survive for a million-plus years.

There is also the fact that we observe primitive tribal cultures in our era that do not pursue invention generation after generation and yet are obviously humans just like us, made in the image of God, who can speak and reason and such. If such tribal cultures exist today, how much more might they have existed before humans enjoyed even the invention of fire-starting, and existed in small tribal groups extremely spread out one from another, each struggling just to survive, with precious little time or spare energy to pursue cultural innovations. Another reason I opted for Piotr Lenartowicz's scenario for an early Adam and Eve is that it is implausible that animals as relatively weak and slow as bipedal *Homo* could have survived without the higher faculties of the soul. Without the capacity for reason, sophisticated verbal communication, foresight, and planning, the world's most dangerous predator would instead be one of its most vulnerable prey.

Equally implausible is the idea of a non-intelligent *Homo* couple giving birth to and rearing Adam. Yes, Pius XII does allow Catholics to consider this to be worthy of serious consideration, so some might regard me as more "Catholic than the Pope" in steering away from it; yet, having seriously considered it, it strikes me as implausible. Did the newborn Adam get a sister Eve, whom he later mated with? If so, what happens to the rich theology of the body which borrows so heavily on Eve's being taken from the side of Adam? How do the parents of this Adam and Eve teach their children language and abstract thought? What happens when Adam and Eve recognize that their parents are mere animals? Is there not something amiss, even disordered, in all this? In any event, it would have necessitated so many exceptional interventions by God that one wonders what the theistic evolutionists gain from opting for this scenario over the much cleaner scenario of Adam created from dust by God and Eve taken from his side.

I remain open to the possibility of a much more recent Adam and Eve, although I think that our first parents were the parents of all the major members of the genus *Homo*, especially if one takes into account the evidence that various *Homo* ancestors mated with one another (with the likely exception of *Homo habilis and Homo naledi*, arguably misgrouped among *Homo*[1]), and that we carry in us the genetic evidence of these fertile trysts. Perhaps I should take more seriously the evidence for the great leap forward a few tens of thousands of years ago, but perhaps those who are focused on this cultural and technological flowering should take more seriously the many other avenues of evidence as to when Adam and Eve lived.

Appendix C.
Faith of Our Fathers— A Hermeneutic of Continuity

Several readers of a draft of this book have raised objections that I think can best be answered in an appendix. And I think that the key to all of them is the importance of continuity to the Christian Faith. Although the comments are at first sight disparate, they are all suggestions that, I believe, lead to an unnecessary rupture with teachings of the Faith. Were they to be accepted, they would not be legitimate developments of the Faith, but distortions. They would make the teachings of the Church unrecognizable to those who taught it and died for it in the early centuries of the Church.

Christ could point to the intricacies and beauty of lilies of the field as evidence of God's care for his creation generally and for man specifically. We can now wonder at the intricacy and the beauty of biological molecules. Moreover, science arose in a Christian culture that believed that God had created the universe according to a rational plan and gave us the rational wherewithal to understand some of its workings. The intelligent design framework is not a science stopper. It is a science starter.

Some people see a greater sign of God's intelligence in a scenario where there is a continuity from physics to abiogenesis to the evolution of all life—past and present. I do not, for three reasons. First,

I don't think that there is evidence for it. Second, I think that the consequence of holding such a view is to deny that God ever acts on creation in a way that science can detect. I have already argued that creation must be permeable to the acts of God—at least insofar as He immediately creates each human soul—and also the laws of creation must allow free beings to be truly free, not constrained by some doctrinaire adherence to a closed universe that must be fully described by physics. There must be space for freedom. Such space is not a figment of the imagination and placeholder for ignorance but rather a feature of the universe. Third, if nature is to be sacramental, somehow reflecting God's plan, then it cannot be left to its own devices to come up with life-forms that surprise God. That would make God learn from nature rather than use nature to teach us.

I have been questioned as to why I am fixated on monogenism. Could we not have an Adam and Eve in one part of the world and perhaps John and Mary in another part? First of all, this is not what Scripture and Tradition and the constant Magisterium of the Church say. A few moments of thought reveal other problems. Adam and Eve have murderous offspring. In no time, they would reproduce and go on a war of conquest against the peace-loving John and Mary and their offspring who did not rebel against God. Moreover, Scripture tells us that it is not just Adam and Eve and their offspring that fell, but the whole of creation. So how does one keep zones of non-fallen nature and fallen nature on the same planet?

One aspect of Scripture that makes it so powerful is that it is grounded in history. Myth is usually timeless and takes place in some make-believe land. The Scriptures on the other hand speak of history. To be sure, the historical presentation takes on different forms in different books of the Bible. The creation of the world is not written in the same genre as the exodus from Egypt, the exploits of King David, or the birth of Christ. But they all are grounded in events that took place on this earth at particular times. They recount a history that the sincere seeker of truth with a modicum of common sense can understand. Certain hyper-intellectuals at times wanted more understanding and built up various schemes of numbered heavens to explain events such as the Ascension of Christ and the Assumption of Mary. But these have just become a stumbling block to later generations and were never part of the original deposit of Faith.

The Faith tells us that the heavens reveal the glory of God. We can argue from the world around us and within us to the existence of God. The early pioneers of science, such as Galileo, spoke of the two books that reveal God's grandeur—the Bible and the book of nature. These two books do not contradict one another. Their teaching is accessible to all thinking human beings who read them. Instead of looking to the materialist scientists whose materialist dogma demands that they ignore scientific evidence contrary to their materialism, we should have the courage, and the humility, to hold on to what God has revealed.

Endnotes

1. Introduction: Wrestling with Darwin

1. Charles Darwin to G. H. Darwin, October 21, 1873, *Darwin Correspondence Project*, Letter no. 9105, University of Cambridge, https://www.darwinproject.ac.uk/letter/?docId=letters/DCP-LETT-9105.xml.
2. George Sim Johnston, "The Genesis Controversy: Darwin's Theory of Evolution Is Losing Support in the Scientific Community," *Crisis*, May 1, 1989, https://www.crisismagazine.com/1989/the-genesis-controversy-darwins-theory-of-evolution-is-losing-support-in-the-scientific-community.
3. Darwin to G. H. Darwin, October 21, 1873.
4. Thomas Henry Huxley, "Criticisms on 'The Origin of Species'" [1964], in *Darwiniana: Essays, Vol. 2* (London: Macmillan and Co., 1893), 147, https://www.gutenberg.org/cache/epub/6919/pg6919-images.html#c3.
5. Huxley, preface to *Darwiniana*, vi.
6. Richard Dawkins, *The Blind Watchmaker* (New York: W.W. Norton, 2006), 10; the first edition of this work was published in 1986.
7. "What Is Evolution?," *BioLogos*, January 15, 2019, https://biologos.org/common-questions/what-is-evolution.
8. Jerry A. Fodor and Massimo Piattelli-Palmarini, *What Darwin Got Wrong* (New York: Picador/Farrar, Straus and Giroux, 2011), xv–xvi.
9. John Tagliabue, "Pope Bolsters Church's Support for Scientific View of Evolution," *New York Times*, October 25, 1996, https://www.nytimes.com/1996/10/25/world/pope-bolsters-church-s-support-for-scientific-view-of-evolution.html.
10. John Shelby Spong, "An Interview with John Shelby Spong: 'I Am Very Orthodox after All!,'" interview by Scott Stephen, *Faith and Theology*, September 5, 2007, https://www.faith-theology.com/2007/09/interview-with-john-shelby-spong-i-am.html.
11. Sergio Almécija et al., "Fossil Apes and Human Evolution," *Science* 372, no. 6542 (May 7, 2021), https://www.science.org/doi/10.1126/science.abb4363.
12. American Museum of Natural History, "Review: Most Human Origins Stories Are Not Compatible with Known Fossils," *Phys.org*, May 6, 2021, https://phys.org/news/2021-05-human-stories-compatible-fossils.html.
13. American Museum of Natural History, "Review: Most Human Origins Stories."

14. Neil Thomas, *Taking Leave of Darwin: A Longtime Agnostic Discovers the Case for Design* (Seattle, WA: Discovery Institute Press), 118–119.
15. Robert F. Shedinger, *The Mystery of Evolutionary Mechanisms* (Eugene, OR: Cascade, 2019), 12.
16. Ernst Mayr, *Populations, Species, and Evolution* (Cambridge, MA: Harvard University Press, 1970), quoted in Shedinger, *The Mystery of Evolutionary Mechanisms*, 3.
17. James A. Shapiro, "In the Details... What?," *National Review*, September 16, 1996, 62–65, https://shapiro.bsd.uchicago.edu/Shapiro.1996.Nat%27lReview.pdf.

2. Evolution: More than a Hypothesis

1. The original French read: "Aujourd'hui, près d'un demi-siècle après la parution de l'encyclique, de nouvelles connaissances conduisent à reconnaître dans la théorie de l'évolution plus qu'une hypothèse." The Vatican website does not provide an authoritative English translation. The text as given on the Pontifical Academy of Sciences website reads: "Today, almost half a century after the publication of the Encyclical, new knowledge has led to the recognition of more than one hypothesis in the theory of evolution." I have chosen the text as it was debated at the time with the problematic phrase "more than a hypothesis." This text and the history of its translation can be found at "Message to the Pontifical Academy of Science on Evolution," *EWTN Global Catholic Network*, last accessed September 25, 2024, https://www.ewtn.com/catholicism/library/message-to-the-pontifical-academy-of-science-on-evolution-8825. It is the text that anyone with a memory of the events should recognize.
2. Giuseppe Sermonti, *Why Is a Fly Not a Horse?* (Seattle, WA: Discovery Institute Press, 2005), 152.
3. See Etienne Gilson, *The Christian Philosophy of St. Augustine* (New York: Random House, 1960), 206.
4. "Maillet, Benoît de (c. 1656–1738)," *Encyclopedia of Philosophy*, last accessed May 6, 2023, https://www.encyclopedia.com/humanities/encyclopedias-almanacs-transcripts-and-maps/maillet-benoit-de-c-1656-1738.
5. Eleanor Marshall, "160th Anniversary of the Presentation of "On the Tendency of Species to Form Varieties," *The Linnean Society*, July 1, 2018, https://www.linnean.org/news/2018/07/01/1st-july-2018-160th-anniversary-of-the-presentation-of-on-the-tendency-of-species-to-form-varieties.
6. Jay W. Richards, ed., introduction to *God and Evolution* (Seattle, WA: Discovery Institute Press, 2010), 7–30; Stephen C. Meyer and Michael Newton Keas, "The Meanings of Evolution," *Stephen C. Meyer*, May 16, 2001, https://stephencmeyer.org/2001/05/16/the-meanings-of-evolution/.
7. Thomas Huxley, preface to *Darwiniana: Essays, Vol. 2* (London: Macmillan and Co., 1893), https://www.gutenberg.org/cache/epub/6919/pg6919-images.html#c1.
8. See Motoo Kimura, "Evolutionary Rate at the Molecular Level," *Nature* 217 (1968): 624–626 and Motoo Kimura, *The Neutral Theory of Molecular Evolution* (Cambridge, UK: Cambridge University Press, 1968). The idea has been further developed by others, including Michael Lynch (e.g., in "The Origins of Eukaryotic

Gee Structure," *Molecular Biology and Evolution* 23 (2006): 450–468. Stephen Meyer critiques this alternate model of evolution in *Darwin's Doubt: The Explosive Origin of Animal Life and the Case for Intelligent Design* (New York: HarperOne, 2013), 321–329.

9. A good discussion of this can be found in James Le Fanu, *Why Us?* (London: Harper Collins, 2009), 111–116.

10. Quoted in Robert F. Shedinger, *The Mystery of Evolutionary Mechanisms* (Eugene, OR: Cascade, 2019), 16.

11. Julian Huxley, "The Evolutionary Vision," in *Evolution After Darwin*, eds. Sol Tax and Charles Callender (Chicago, IL: Chicago University Press, 1960), 252–253, 260. The University of Chicago Darwin Centennial Celebration took place November 24–28, 1959.

12. Stephen Jay Gould, "Impeaching a Self-Appointed Judge," *Scientific American* (July 1992): 118–121.

13. David L. Hull, "The God of the Galápagos," *Nature* 352 (August 8, 1991): 485–486.

14. See Stephen C. Meyer, *Darwin's Doubt: The Explosive Origin of Animal Life and the Case for Intelligent Design* (New York: Harper Collins, 2013), 72.

15. Meyer, *Darwin's Doubt*, 77–79.

16. Simon Conway Morris, "The Cambrian 'Explosion': Slow-Fuse or Megatonnage?," *PNAS* 97, no. 9 (April 25, 2000): 4426–4429, https://doi.org/10.1073/pnas.97.9.4426.

17. Meyer, *Darwin's Doubt*, 32.

18. Meyer, *Darwin's Doubt*, 33.

19. Jerry A. Fodor and Massimo Piattelli-Palmarini, *What Darwin Got Wrong* (New York: Picador/Farrar, Straus and Giroux, 2011), 52.

20. Wells, quoted in Meyer, *Darwin's Doubt*, 97.

21. Günter Bechly and Stephen C. Meyer, "The Fossil Record and Universal Common Ancestry," in *Theistic Evolution*, ed. J. P. Moreland, et al. (Wheaton, IL: Crossway, 2017).

22. Stephen Jay Gould, "Evolution's Erratic Pace," *Natural History* 86 (May 1987): 14.

23. Sermonti, *Why Is a Fly Not a Horse?*, 146.

24. John A. Davison, *The Unpublished Evolutionary Papers of John A. Davison*, ed. Phillip L. Engle (Greensburg, PA: Laurel Highlands Media, 2000), 146.

25. Stephen Jay Gould, *The Structure of Evolutionary Theory* (Cambridge, MA: Belknap, 2002), 452.

26. Richard Goldschmidt, *The Material Basis of Evolution* (New Haven, CT: Yale University Press, 1940), 6–7.

27. Charles Darwin, *On the Origin of the Species by Means of Natural Selection* (London: John Murray, 1859), chap. 6, "Difficulties on Theory," subsection "Organs of extreme perfection and complication."

28. See Meyer, *Darwin's Doubt*, 169–184.

29. Paul Nelson, "50 Years of Scientific Challenges to Evolution: Remembering the Wistar Symposium," *YouTube*, April 25, 2016, video, 6:03, https://www.youtube

.com/watch?v=VQ_y12X_Sm2k. See also David Klinghoffer, "For Darwin Advocates, Wistar Conference Remains a Pain in the Master Narrative," *Evolution News & Science Today*, April 25, 2016, https://evolutionnews.org/2016/04/for_darwin_advo/. Also see *Mathematical Challenges to the Neo-Darwinian Interpretation of Evolution* (Philadelphia, PA: Wistar Institute Press, 1967), 80.
30. See Mathematical Challenges to the Neo-Darwinian Interpretation of Evolution (Philadelphia, PA: Wistar Institute Press, 1967), 80.
31. Douglas D. Axe, "Estimating the Prevalence of Protein Sequences Adopting Functional Enzyme Folds," *Journal of Molecular Biology* 341, no. 5 (August 27, 2004), 1295–1315. For other research corroborating Axe's findings, see Stephen C. Meyer, *Return of the God Hypothesis: Three Scientific Discoveries That Reveal the Mind Behind the Universe* (New York: HarperOne, 2021), 319–322 and 482n18. There are what are referred to as "intrinsically disordered proteins," many of which, according to Axe, would be better described as "conditionally folded proteins." It has been argued that the existence of such proteins poses a challenge to Axe's argument. He rebuts this and other objections in his essay "Losing the Forest by Fixating on the Trees—A Response to Venema's Critique of *Undeniable*," *Evolution News & Science Today*, February 6, 2018, https://evolutionnews.org/2018/02/losing-the-forest-by-fixating-on-the-trees-a-response-to-venemas-critique-of-undeniable/.
32. Pengfei Tian and Robert B. Best, "How Many Protein Sequences Fold to a Given Structure? A Coevolutionary Analysis," *Biophysics Journal*, 113, no. 8 (October 17, 2017), 1719–1730, DOI: 10.1016/j.bpj.2017.08.039.
33. See Michael Behe, *Edge of Evolution* (New York: Free Press), 63.
34. "Weasel Program," *Wikipedia*, September 3, 2023, https://en.wikipedia.org/wiki/Weasel_program#Implications_for_biology. Dawkins's illustration involves a prespecification, to use Dembski's language in *The Design Inference*. Dembski also develops the theoretical apparatus for handling specifications that are not prespecified, such as the odds of arriving at any meaningful/grammatical/functional sequence of a given length of letters. Such a case is, of course, harder to calculate, but is still extraordinarily unlikely when dealing with text strings of a length similar to the Methinks sentence from *Hamlet*. See further down in this chapter where I discuss what research in molecular biology is revealing about how common functional ("meaningful") amino acid sequences (ones that lead to functional proteins) are amidst the set of all possible amino acid sequences of a given length.
35. Fred Hoyle, "The Steady State Theory Revised," *Comments on Astrophysics*, 13, no. 2 (1989), 85.
36. E. V. Koonin, *The Logic of Chance: The Nature and Origin of Biological Evolution* (Upper Saddle River, NJ: Pearson Education, 2012), 435.
37. Michael Behe, *Darwin Devolves* (New York: HarperOne, 2019), 104–106 and 133.
38. Mark Pallen and Nicholas Matzke, "From *The Origin of Species* to the Origin of Bacterial Flagella," *Nature Reviews Microbiology* 4 (2006), 784–790, https://doi.org/10.1038/nrmicro1493.
39. Jonathan McLatchie, "Michael Behe Hasn't Been Refuted on the Flagellum," March 15, 2011, *Evolution News & Science Today*, https://evolutionnews.org/2011/03/michael_behe_hasnt_been_refute/.

40. On Miller's approbation, see Jennifer Cutraro, "A Complex Tail, Simply Told," *Science*, April 17, 2007, https://www.science.org/content/article/complex-tail-simply-told.
41. N. Matzke, "Flagellum Evolution Paper Exhibits Canine Qualities," *Panda's Thumb*, April 16, 2007, https://pandasthumb.org/archives/2007/04/flagellum-evolu-1.html.
42. See Behe, *Darwin Devolves*, 293.
43. Franklin M. Harold, *The Way of the Cell* (New York: Oxford University Press, 2001), 205. See Michael Behe, "Best of Behe: Blind Evolution or Intelligent Design? An Address at the American Museum of Natural History," *Evolution News & Science Today*, December 18, 2016, https://evolutionnews.org/2016/12/best_of_behe_bl/.
44. Michael Behe, *Darwin's Black Box* (New York: Simon and Schuster, 1996), 179–183.
45. Wojciech Makalowski, "What Is Junk DNA, and What Is It Worth?," *Scientific American* (February 12, 2007), https://www.scientificamerican.com/article/what-is-junk-dna-and-what/.
46. See, for example, William Dembski, "Intelligent Science and Design," *First Things* 86 (October 1998): 21–27. Prior to Dembski's 1998 article, in 1994, ID proponent Forrest Mims wrote a letter to *Science* in which he he said there might be use for the "junk" DNA, but the letter wasn't published, not surprising since Mims's prediction flew in the face of Darwinian orthodoxy. See Casey Luskin, "New *Long Story* Video Tackles "A Battle of Predictions: Junk DNA,'" *Evolution News & Science Today*, March 28, 2024, https://evolutionnews.org/2024/03/new-long-story-video-tackles-a-battle-of-predictions-junk-dna/.
47. Encode Project Consortium, "An Integrated Encyclopedia of DNA Elements in the Human Genome," *Nature* 489 (September 5, 2012): 57–74, https://www.nature.com/articles/nature11247.
48. Francis Collins, speech, 33rd Annual J. P. Morgan Healthcare Conference, San Francisco, CA, January 13, 2015. Quoted in Marvin Olasky, "Admission of Function: Junking Slurs about 'Junk DNA,'" *WORLD*, July 9, 2016, https://wng.org/articles/admission-of-function-1620609700.
49. Michael Denton, *Evolution: Still a Theory in Crisis* (Seattle, WA: Discovery Institute Press, 2016), 83.
50. Denton, *Evolution: Still a Theory in Crisis*, 83.
51. Sermonti, *Why Is a Fly Not a Horse?*, 68.
52. Ernst Mayr, *Animal Species and Evolution* (Cambridge, MA: Harvard University Press, 1963), 609.
53. Meyer, *Darwin's Doubt*, 271.
54. Sermonti, *Why Is a Fly Not a Horse?*, 104.
55. Sermonti, *Why Is a Fly Not a Horse?*, 107.
56. Sermonti, *Why Is a Fly Not a Horse?*, 110.
57. See "Morphogenesis: Coding for Shape," *Evolution News & Science Today*, October 19, 2020, https://evolutionnews.org/2020/10/morphogenesis-coding-for-shape/.

58. Philip Ball, *How Life Works: A User's Guide to the New Biology* (Chicago, IL: University of Chicago Press, 2024), 94.
59. Denis Noble, "It's Time to Admit That Genes Are Not the Blueprint for Life," review of *How Life Works* by Philip Ball, *Nature* (February 5, 2024), https://www.nature.com/articles/d41586-024-00327-x. See also Casey Luskin, "Denis Noble in *Nature*: 'Time to Admit Genes Are Not the Blueprint For Life,'" *Evolution News & Science Today*, February 16, 2024, https://evolutionnews.org/2024/02/denis-noble-in-nature-time-to-admit-genes-are-not-the-blueprint-for-life/.
60. See Eric Cassell, *Animal Algorithms: Evolution and the Mysterious Origin of Ingenious Instincts* (Seattle, WA: Discovery Institute Press, 2021).
61. Charles Darwin, *On the Origin of Species*, 1st ed. (London: John Murray, 1859), 201, https://darwin-online.org.uk/content/frameset?itemID=F373&viewtype=image&pageseq=1.
62. Wolf-Ekkehard Lönnig, "Plant Galls and Evolution," *Wolf-Ekkehard Lönnig*, September 7, 2017, https://www.weloennig.de/PlantGalls.pdf.
63. Marion O. Harris and Andrea Pitzschke, "Plants Make Galls to Accommodate Foreigners: Some Are Friends, Most Are Foes," *New Phytologist* 225, no. 5 (November 27, 2019): 1853, https://nph.onlinelibrary.wiley.com/doi/epdf/10.1111/nph.16340. This definition is quoted by Lönnig, not in the article just mentioned, but in his "Plant Galls, Evolution, and Intelligent Design," *Evolution News & Science Today*, August 28, 2020, https://evolutionnews.org/2020/08/plant-galls-evolution-and-intelligent-design/.
64. Joachim Illies, quoted in "Plant Galls, Evolution, and Intelligent Design," *Evolution News & Science Today*, August 28, 2020, https://evolutionnews.org/2020/08/plant-galls-evolution-and-intelligent-design/. J. Illies, "Gallenbitteris Ärgernis," *Natur—Horst Sterns Umweltmagazin* 6 (Juni 1981): 42–47.
65. B. D. McKay and R. M. Zink, "Sisyphean Evolution in Darwin's Finches," *Biological Reviews* 90 (2014): 689–698.
66. See Behe, *Darwin Devolves*, 149–152.
67. Behe, *Darwin Devolves*, 161.
68. Behe, *Darwin Devolves*, 162.
69. Brian Miller points to recent peer reviewed research to argue that "cichlid variation did not primarily originate from random mutations but from systems engineered to drive targeted modifications." Miller, "Studies on Cichlid Fish Demonstrate the Predictive Power of Engineering Models for Adaptation," *Evolution News & Science Today*, October 14, 2021, https://evolutionnews.org/2021/10/studies-on-cichlid-fish-demonstrate-the-predictive-power-of-engineering-models-for-adaptation/.
70. Behe, *Darwin Devolves*, 166.
71. "How Hybrids Have Upturned Evolutionary Theory," *Economist*, October 3, 2020, 67–70, https://www.economist.com/science-and-technology/2020/10/03/how-hybrids-have-upturned-evolutionary-theory.
72. Behe, *Darwin Devolves*, 17.
73. "*E. coli* Long-Term Evolution Experiment," *Wikipedia*, September 2, 2024, https://en.wikipedia.org/wiki/E._coli_long-term_evolution_experiment.

74. Matti Leisola and Jonathan Witt, *Heretic: One Scientist's Journey from Darwin to Design* (Seattle, WA: Discovery Institute Press, 2018), loc. 2371ff, Kindle.
75. Michael Behe, "Citrate Death Spiral," *Evolution News & Science Today*, June 17, 2020, https://evolutionnews.org/2020/06/citrate-death-spiral/.
76. J. Grey Monroe et al., "The Population Genomics of Adaptive Loss of Function," *Nature Heredity* 126 (2021): 383–395. Internal citations removed. As quoted in Casey Luskin, "Vindicated but Not Cited: Paper in *Nature Heredity* Supports Michael Behe's Devolution Hypothesis," *Evolution News & Science Today*, February 16, 2021, https://evolutionnews.org/2021/02/vindicated-but-not-cited-paper-in-nature-heredity-supports-michael-behes-devolution-hypothesis/.
77. Michael J. Behe, "Experimental Evolution, Loss-of-Function Mutations, and the 'First Rule of Adaptive Evolution,'" *Quarterly Review of Biology* 85, no. 4 (December 2010):419–445, https://pubmed.ncbi.nlm.nih.gov/21243963/.
78. See Andrzej Elzanowski and Jim Ostell, "The Genetic Codes," *NCBI Taxonomy Broswer*, September 23, 2024, https://www.ncbi.nlm.nih.gov/Taxonomy/taxonomyhome.html/index.cgi?chapter=cgencodes.
79. Stephen Freeland et al., "Early Fixation of an Optimal Genetic Code," *Molecular Biology and Evolution* 17, no. 4 (May 2000): 511–518, https://www.researchgate.net/publication/12574359_Early_Fixation_of_an_Optimal_Genetic_Code; Shalev Itzkovitz and Uri Alon, "The Genetic Code Is Nearly Optimal for Allowing Additional Information within Protein-Coding Sequences," *Genome Research* 17, no. 4 (2007): 405–412, https://genome.cshlp.org/content/17/4/405; and Małgorzata Wnętrzak, Paweł Błażej, Paweł Mackiewicz, "Optimization of the Standard Genetic Code in Terms of Two Mutation Types: Point Mutations and Frameshifts," *Biosystems* 181 (July 2019): 44–50, https://www.sciencedirect.com/science/article/abs/pii/S0303264719301170?via%3Dihub.
80. Casey Luskin, "Problem 6: Molecular Biology Has Failed to Yield a Grand 'Tree of Life,'" *Evolution News & Science Today*, February 15, 2015, https://evolutionnews.org/2015/02/problem_6_molec/.
81. Liliana M. Dávalos et al. "Understanding phylogenetic incongruence: lessons from phyllostomid bats," Biological Reviews of the Cambridge Philosophical Society, Vol. 87:991-1024 (2012).
82. Leonidas Salichos and Antonis Rokas, "Inferring Ancient Divergences Requires Genes with Strong Phylogenetic Signals," *Nature* 497 (May 8, 2013): 327–331, https://www.nature.com/articles/nature12130.
83. Perry Marshall, *Evolution 2.0: Breaking the Deadlock Between Darwin and Design* (Dallas, TX: Benbella, 2015), 124.
84. See Casey Luskin, "Problem 9: Neo-Darwinism Struggles to Explain the Biogeographical Distribution of Many Species," *Evolution News & Science Today*, February 16, 2015, https://evolutionnews.org/2015/02/problem_9_neo-d/#fn153.
85. I depend on Wells's discussion of this topic found in his *Icons of Evolution: Why Much of What We Teach about Evolution is Wrong* (Washington, DC: Regnery, 2000), 81–109.
86. Charles Darwin to Asa Gray, September 10, 1860, *Darwin Correspondence Project*, Letter no. 2910, https://www.darwinproject.ac.uk/letter/?docId=letters/DCP-LETT-2910.xml.

87. For examples of biology textbooks featuring this discredited icon of evolution, see Chapter 5 of Jonathan Wells, *Icons of Evolution*. See also Jonathan Wells, *Zombie Science: More Icons of Evolution* (Seattle, WA: Discovery Institute Press, 2017), 59–60.
88. See Casey Luskin, "The Textbooks Don't Lie: Haeckel's Faked Drawings *Have* Been Used to Promote Evolution: Miller & Levine (1994) (Part I)," *Evolution News & Science Today*, May 26, 2007, https://evolutionnews.org/2007/05/the_textbooks_dont_lie_haeckel/.
89. See Cornelius Hunter, *Darwin's God* (Grand Rapids, MI: Brazos, 2001), 31–33.
90. Sermonti, *Why Is a Fly Not a Horse?*, 11–12.
91. Massimo Piattelli-Palmarini, quoted in Suzan Mazur, *The Altenberg 16: An Exposé of the Evolution Industry* (Berkeley, CA: North Atlantic Books, 2009), 317.
92. See Sermonti, *Why Is a Fly Not a Horse?*, 25.
93. Charles Darwin to J. D. Hooker, February 1 [1871], *Darwin Correspondence Project*, Letter no. 7471, University of Cambridge, https://www.darwinproject.ac.uk/letter/?docId=letters/DCP-LETT-7471.
94. Thomas Aquinas, *Summa Theologiae* [1274] I, question 73, article 1, objection 3. See *The Summa Theologiæ of St. Thomas Aquinas,* trans. Fathers of the English Dominican Province [1920], https://www.newadvent.org/summa/1073.htm.
95. See David Wooton, *Bad Medicine: Doctors Doing Harm Since Hippocrates* (Oxford, UK: Oxford University Press, 2006), 132.
96. Thomas Henry Huxley, "On Some Organisms Living at Great Depths in the North Atlantic Ocean," *Quarterly Journal of Microscopical Science* 8 (1868): 205.
97. Philip F. Rehbock, "Huxley, Haeckel, and the Oceanographers: The Case of *Bathybius haeckelii*," *Isis* 66, no. 4 (December 1975): 505.
98. Dean Kenyon, foreword to *The Mystery of Life's Origin: The Continuing Controversy*, by Charles B. Thaxton et al. (Seattle, WA: Discovery Institute Press, 2020), 39.
99. James Tour, "We're Still Clueless about the Origin of Life," in *The Mystery of Life's Origin*, 324. Excerpted at Discovery Institute, https://www.discovery.org/a/were-still-clueless-about-the-origin-of-life/.
100. Tour, "We're Still Clueless about the Origin of Life," 325.
101. James Tour, "Can Scientists Answer These Questions? RNA, Abiogenesis, Chemical Natural Selection & More," *YouTube*, August 23, 2023, video, 23:53, https://youtu.be/MmykRoelTzU?si=2_AH9JF0OwPL5Yye.
102. Michael Marshall, "The Water Paradox and the Origins of Life," *Nature* 588 (December 10, 2020): 210–213, https://media.nature.com/original/magazine-assets/d41586-020-03461-4/d41586-020-03461-4.pdf.
103. Nick Lane and Joana C. Xavier, "To Unravel the Origin of Life, Treat Findings as Pieces of a Bigger Puzzle," *Nature* 626 (February 26, 2024): 948–951.
104. David Coppedge, "From *Nature*, a Devastating Critique of Origin-of-Life Research," *Evolution News & Science Today*, February 28, 2024, https://evolutionnews.org/2024/02/from-nature-a-devastating-critique-of-origin-of-life-research/.
105. Mark Ridley, *The Problems of Evolution* (Oxford: Oxford University Press, 1985), 35.

3. Intelligent Design in Nature

1. Antony Flew, *There Is a God: How the World's Most Notorious Atheist Changed His Mind* (New York: Harper Collins, 2007), 132.
2. Henri Bergson, *Creative Evolution*, trans. Arthur Mitchell (New York: Henry Holt and Co., 1911), 335.
3. Isaac Newton, *Principia in Modern English: Isaac Newton's Mathematical Principles of Natural Philosophy & His System of the World*, trans. Andrew Motte, revised by Florian Cajori (Berkeley, CA: University of California Press, 1934), 398–400.
4. Joseph-Louis Lagrange, quoted by Forest Ray Moulton, *Introduction to Astronomy* (New York: Macmillan, 1916), 234, https://archive.org/details/introductioastro00moulrich/page/234/mode/2up.
5. Isaac Newton, *Newton's Principia: The Mathematical Principle of Natural Philosophy*, trans. Andrew Motte (New York: Daniel Adee, 1846), 385, https://archive.org/embed/newtonspmathema00newtrich.
6. To be sure, in real life, being wedded to a theory can influence our reading and presentation of the data. And this seems to have happened to a certain extent with the 1919 solar eclipse expedition to test Einstein's theory. But there is always the possibility of doing the test again, as was done in the case of Einstein's theory of general relativity.
7. Richard Lewontin, "Billions and Billions of Demons," review of *The Demon-Haunted World: Science as a Candle in the Dark*, by Carl Sagan, *The New York Review*, January 9, 1997, 31.
8. Phillip E. Johnson, "Shouting Heresy in the Temple of Darwin," *Christianity Today*, October 24, 1994, https://www.christianitytoday.com/1994/10/article-shouting-heresy-in-temple-of-darwin/.
9. Paul A. Nelson, "Life in the Big Tent: Traditional Creationism and the Intelligent Design Community," *Christian Research Institute*, June 9, 2009, https://www.equip.org/articles/life-in-the-big-tent/.
10. "How Many Organisms Have Ever Lived on Earth?," *Biology StackExchange*, last accessed September 26, 2024, https://biology.stackexchange.com/questions/6937/how-many-organisms-have-ever-lived-on-earth.
11. William Dembski and Jonathan Witt, *Intelligent Design Uncensored: An Easy-to-Understand Guide to the Controversy* (Downers Grove, IL: InterVarsity Press, 2010), 68–69.
12. See Edward Feser, "'Intelligent Design' Theory and Mechanism," *Edward Feser*, April 10, 2010, https://edwardfeser.blogspot.com/2010/04/intelligent-design-theory-and-mechanism.html.
13. See "Morphogenesis: Coding for Shape," *Evolution News & Science Today*, October 19, 2020, https://evolutionnews.org/2020/10/morphogenesis-coding-for-shape/.
14. For a partial list, see Jay Richards, "List of Fine-Tuning Parameters," *Discovery Institute*, January 14, 2015, https://www.discovery.org/a/fine-tuning-parameters/.
15. See Roger Penrose, *The Road to Reality: A Complete Guide to the Laws of the Universe* (London: Jonathan Cape, 2004), 729–731; and Roger Penrose, *The Emperor's*

New Mind: Concerning Computers, Minds, and the Laws of Physics (New York: Oxford University Press, 1990), 444–445.

16. See, for example, "Is the Universe Fine Tuned for Life? Sir Roger Penrose vs William Lane Craig," Premiere Unbelievable?, *YouTube*, October 30, 2019, video, 10:01, https://youtu.be/OBAbjE-WOJo?si=9PuSI-r_TXu-5JCQ.
17. Richards, "List of Fine-Tuning Parameters."
18. Kenneth R. Miller, *Finding Darwin's God* (New York: Harper Collins, 1999), 231.
19. Denise Kirschner, Mark Chaplain, and Akira Sasaki, "Disclaimer," *Journal of Theoretical Biology* 506 (December 7, 2020), https://doi.org/10.1016/j.jtbi.2020.110456.
20. "Censor of the Year: Wikipedia," *Free Science*, 2018, https://freescience.today/censor/.
21. Larry Sanger, quoted in David Klinghoffer, "Wikipedia Co-Founder Blasts 'Appallingly Biased' Wikipedia Entry on Intelligent Design," *Evolution News & Science Today*, December 12, 2017, https://evolutionnews.org/2017/12/wikipedia-co-founder-calls-wikipedia-entry-on-intelligent-design-appallingly-biased/.
22. "Stories," *Free Science*, last accessed September 16, 2024, https://freescience.today/stories/.
23. Jonathan Wells, "'In China We Can Criticize Darwin': Prelude," *Evolution News & Science Today*, April 16, 2014, https://evolutionnews.org/2014/04/in_china_we_can/.
24. Eric Hedin, *Canceled Science: What Some Atheists Don't Want You to See* (Seattle, WA: Discovery Institute Press, 2021), 11.
25. Laurence A. Moran, "Inside Higher Ed Weighs in on the Ball State Academic Freedom Controversy," *Sandwalk: Strolling with a Skeptical Biochemist*, May 17, 2013, https://sandwalk.blogspot.com/2013/05/inside-higher-ed-weighs-in-on-ball.html.
26. Thomas Nagel, *Mind and Cosmos: Why the Materialist Neo-Darwinian Conception of Nature Is Almost Certainly False* (New York: Oxford University Press, 2012), 10–11.
27. Michael Behe, *A Mousetrap for Darwin: Michael J. Behe Answers His Critics* (Seattle, WA: Discovery Institute Press, 2020), 347–355.
28. Michael Behe, *The Edge of Evolution: The Search for the Limits of Darwinism* (New York: Free Press), 201.
29. Behe, *Mousetrap*, 215–216.
30. See Paul Nelson, "Our 20th- and 21st-Century Ptolemaic Epicycles'?," *Evolution News & Science Today*, November 30, 2020, https://evolutionnews.org/2020/11/our-20th-and-21st-century-ptolemaic-epicycles/.
31. C. John Collins, *Genesis 1–4: A Linguistic, Literary, and Theological Commentary* (Phillipsburg, NJ: P&R Publishing, 2006), 44.
32. Stanley L. Jaki, "Genesis 1: A Cosmogenesis?," *Homiletic & Pastoral Review*, August 1, 1993, https://www.hprweb.com/1993/08/genesis-1-a-cosmogenesis/.
33. David R. Montgomery, *The Rocks Don't Lie: A Geologist Investigates Noah's Flood* (New York: Norton, 2013), 61–63.
34. All references to the Bible are from the Revised Standard Version Catholic Edition, unless noted otherwise.

35. See, for example, Michael Behe, "Best of Behe: Blind Evolution or Intelligent Design? An Address at the American Museum of Natural History," *Evolution News & Science Today*, December 18, 2016, https://evolutionnews.org/2016/12/best_of_behe_bl/.
36. Sermonti, *Why Is a Fly Not a Horse?*, 114.
37. Pius XII, *Humani Generis*, encyclical letter, Vatican, August 12, 1950, 36, https://www.vatican.va/content/pius-xii/en/encyclicals/documents/hf_p-xii_enc_12081950_humani-generis.html.
38. Thomas Aquinas, *Summa Theologiae* [1274] I, question 75, article 2; I, question 90, article 2; *De Ente et Essentia (On Being and Essence)*.
39. "List of Peer-Reviewed and Mainstream Scientific Publications Supporting Intelligent Design," *Discovery Institute*, May 2024, https://www.discovery.org/m/securepdfs/2024/05/Peer-Reviewed-and-Mainstream-Articles-Page-Update-May-2024_FinalPDF.pdf.
40. Ernst Robert Curtius, *European Literature and the Latin Middle Ages*, trans. Willard R. Trask (New York: Pantheon, 1953), 504.
41. Whitehead, the son of an Anglican clergyman, maintained his Christian faith in his early years as a student in Cambridge. But "later he moved away from religion, settling his long vacillation between the various Christian churches, for agnosticism or even the atheistic materialism of his friends, the first being Bertrand Russell." Rémy Lestienne, *Alfred North Whitehead: Philosopher of Time* (New Jersey: World Scientific, 2022), chap. 11, "Whitehead's God," 195–221, https://www.worldscientific.com/doi/10.1142/9781800611788_0011.
42. Alfred North Whitehead, *Science and the Modern World* (New York: Free Press, 1967), 12.
43. Robert Laughlin, quoted in John Lennox, *Seven Days that Divide the World* (Grand Rapids, MI: Zondervan, 2011), 183.
44. See David Klinghoffer, "Michael Denton and Intelligent Design's Big Tent," *Evolution News & Science Today*, January 20, 2020, https://evolutionnews.org/2020/01/denton-and-intelligent-designs-big-tent/.
45. Nagel, *Mind and Cosmos*, 12.
46. Thomas Nagel, *The Last Word* (Oxford, UK: Oxford University Press, 1997), 130.

4. Intelligent Design: A Preamble to a Powerful Way to God

1. Charles Darwin to Asa Gray, April 3, 1860, *Darwin Correspondence Project*, Letter no. 2743, https://www.darwinproject.ac.uk/letter/?docId=letters/DCP-LETT-2743.xml.
2. Pius IX, *Dogmatic Constitution on the Catholic Faith: Dei Filius* (Vatican I), April 24, 1870, chap. 2, https://www.catholicplanet.org/councils/20-Dei-Filius.htm.
3. John Paul II, *Catechism of the Catholic Church* [1993], 2nd ed., (Washington DC: United States Catholic Conference, 2011), para. 31, https://www.usccb.org/sites/default/files/flipbooks/catechism/.

4. Aquinas quoted in Michael Chaberek, *Aquinas and Evolution: Why St. Thomas's Teachings on the Origins is Incompatible with Evolutionary Theory* (British Columbia, Canada: Chartwell Press, 2019), 171. Thomas Aquinas, *Questiones Disputatae de Veritate* [1256–1259], trans. Robert W. Mulligan (Chicago, IL: Henry Regnery, 1951), question 5, "Providence," article 2, "Is the World Ruled by Providence?" reply, https://isidore.co/aquinas/english/QDdeVer5.htm#5.
5. Aquinas quoted in Chaberek, *Aquinas*, 171. Thomas Aquinas, *Summa Contra Gentiles, Book III: Providence* [c. 1263], trans. Vernon J. Bourke (Notre Dame, IN: University of Notre Dame Press, 1956), chap. 3, para. 9, https://classicalliberalarts.com/st-thomas-aquinas-summa-contra-gentiles/st-thomas-aquinas-summa-contra-gentiles-book-iii-of-providence/#Chapter-3.--That-Every-Agent-Acts-for-a-Good.
6. See Tim Folger, "Science's Alternative to an Intelligent Creator: The Multiverse Theory," *Discover*, November 10, 2008, https://www.discovermagazine.com/the-sciences/sciences-alternative-to-an-intelligent-creator-the-multiverse-theory.
7. Many, but not all, Catholic and Orthodox Bibles follow the Septuagint's numbering of the Psalms, in which Psalms 11–113 in Hebrew are Psalms 10–112 in the Septuagint. The references here follow the numbering of the Hebrew Psalms.
8. John G. West, "Nothing New Under the Sun," in *God and Evolution*, ed. Jay Richards (Seattle, WA: Discovery Institute, 2010), 35–36.
9. Immanuel Kant, *Critique of Pure Reason*, trans. Paul Guyer and Allen W. Wood (Cambridge, UK: Cambridge University Press, 1998), 579.
10. Roger Scruton, *Kant* (Oxford, UK: Oxford University Press, 1982), 52.
11. John Paul II, "The Proofs for God's Existence," General Audience, July 10, 1985. The English translation cited here is from https://inters.org/John-Paul-II-Science-Proofs-God.
12. Pat Flynn, "Two Catholic Scientists Debate Intelligent Design," *Chronicles of Strength*, June 2019, https://www.chroniclesofstrength.com/two-catholic-scientists-debate-intelligent-design/.
13. See Behe's talk, delivered at the American Museum of Natural History April 23, 2002, reproduced in "Best of Behe: Blind Evolution or Intelligent Design? An Address at the American Museum of Natural History," *Evolution News & Science Today*, December 18, 2016, https://evolutionnews.org/2016/12/best_of_behe_bl/.
14. John Paul II, *Catechism of the Catholic Church* [1993], 2nd ed. (Washington DC: United States Catholic Conference, 2011), para. 33, https://www.usccb.org/sites/default/files/flipbooks/catechism/.
15. John Paul II, "Message to the Pontifical Academy of Science on Evolution," October 22, 1996, section 5. Available at EWTN, https://www.ewtn.com/catholicism/library/message-to-the-pontifical-academy-of-science-on-evolution-8825.
16. Amy Mathews Amos, "Why They Stray: The Evolutionary Advantages of Infidelity," *Pacific Standard*, November 5, 2014, https://psmag.com/social-justice/stray-evolutionary-advantages-infidelity-93443/.
17. Paul W. Andrews and J. Anderson Thomson, Jr., "Depression's Evolutionary Roots," *Scientific American*, August 25, 2009, https://www.scientificamerican.com/article/depressions-evolutionary/.

18. "Why Cry? Evolutionary Biologists Show Crying Can Strengthen Relationships," *Science Daily*, September 7, 2009, https://www.sciencedaily.com/releases/2009/08/090824141045.htm.
19. Silvio José Lemos Vasconcellos et al., "Understanding Lies Based on Evolutionary Psychology: A Critical Review," *Trends in Psychology* 27, no. 1 (March 2019), https://www.scielo.br/scielo.php?pid=S2358-18832019000100141&script=sci_arttext.
20. W. V. O. Quine, "Natural Kinds," in *Essays in Honor of Carl G. Hempel*, ed. N. Rescher (Dordrecht, Netherlands: Reidel, 1969), 13.
21. Jerry A. Fodor and Massimo Piattelli-Palmarini, *What Darwin Got Wrong* (New York: Picador/Farrar, Straus and Giroux, 2011), 194.
22. Karl Popper, *Unended Quest: An Intellectual Biography* (London: Fontana, 1976), 171–172.
23. Charles Darwin, *On the Origin of the Species by Means of Natural Selection* (London: John Murray, 1859), chap. 14, 485, https://darwin-online.org.uk/content/frameset?itemID=F373&viewtype=image&pageseq=503.
24. James Lennox and Charles H. Pence, "Darwinism," *The Stanford Encyclopedia of Philosophy* (Summer 2024), https://plato.stanford.edu/entries/darwinism/.

5. Creation Groans

1. Charles Darwin to Asa Gray, May 22, 1860, *Darwin Correspondence Project*, Letter no. 2814, University of Cambridge, https://www.darwinproject.ac.uk/letter/?docId=letters/DCP-LETT-2814.xm.
2. Cornelius G. Hunter, *Darwin's God: Evolution and the Problem of Evil* (Grand Rapids, MI: Brazos Press, 2001).
3. William Paley, *Natural Theology, or Evidence of the Existence and Attributes of the Deity, Collected from the Appearances of Nature* (London: R. Faulder, 1802), 456.
4. See, for example, Steve Dilley, "Nothing in Biology Makes Sense Except in Light of Theology?," *Studies in History and Philosophy of Science Part C: Studies in History and Philosophy of Biological and Biomedical Sciences* 44, no. 4, part B (December 2013): 774–786.
5. "Praying Mantis Love is Waaay Weirder Than You Think," Deep Think, *YouTube*, November 14, 2017, video, 5:08, https://www.youtube.com/watch?v=NHf47gI8w04.
6. "Why the Male Black Widow Is a Real Home Wrecker," Deep Look, *YouTube*, January 9, 2018, video, 5:48, https://www.youtube.com/watch?v=NpJNeGqExrc. Actually, in most species of the black widow, this cannibalism does not take place.
7. Charles Darwin, *On the Origin of the Species by Means of Natural Selection* (London: John Murray, 1859), chap. 7, https://www.gutenberg.org/files/1228/1228-h/1228-h.htm.
8. Darwin stated this view perhaps most explicitly in an 1881 letter to William Graham. There he wrote, "I could show fight on natural selection having done and doing more for the progress of civilisation than you seem inclined to admit. Remember what risks the nations of Europe ran, not so many centuries ago of being

overwhelmed by the Turks, and how ridiculous such an idea now is. The more civilised so-called Caucasian races have beaten the Turkish hollow in the struggle for existence. Looking to the world at no very distant date, what an endless number of the lower races will have been eliminated by the higher civilised races throughout the world." Charles Darwin to William Graham, July 3, 1881, *Darwin Correspondence Project*, Letter no. 13230, University of Cambridge, https://www.darwinproject.ac.uk/letter/?docId=letters/DCP-LETT-13230.xml. See also Charles Darwin, *The Descent of Man and Selection in Relation to Sex* [1871] (London: John Murray, 1901), Part 1, chap. 6 and chap. 7, https://archive.org/details/ncbs.BB-001_0_0_0_1/mode/2up.

9. See Ed Yong, "Ballistic Penises and Corkscrew Vaginas: The Sexual Battles of Ducks," *Discover*, December 22, 2009, https://www.discovermagazine.com/planet-earth/ballistic-penises-and-corkscrew-vaginas-the-sexual-battles-of-ducks.

10. Richard O. Prum, *The Evolution of Beauty: How Darwin's Forgotten Theory of Mate Choice Shapes the Animal World—and Us* (New York: Doubleday, 2017), 157–158.

11. See Dylan Matthews, "7 Adorable Animals That Are Also Murderous Monsters," *Vox*, March 30, 2016, https://www.vox.com/2015/2/23/8089019/horrible-animals.

12. Dieter Lukas and Elise Huchard, "The Evolution Infanticide by Males in Mammalian Societies," *Science* 346, no. 6211 (November 14, 2014): 841–844, https://www.science.org/doi/10.1126/science.1257226.

13. "Unraveling Why Some Mammals Kill Off Infants," *New York Times*, November 13, 2014, https://www.nytimes.com/2014/11/13/science/unraveling-why-some-mammals-kill-off-infants.html. See also Lukas and Huchard, "The Evolution of Infanticide."

14. Laurence Culot et al., "Reproductive Failure, Possible Maternal Infanticide, and Cannibalism in Wild Moustached Tamarins, *Saguinus mystax*," *Primates* 52, no. 2 (February 17, 2011): 179–186.

15. "Panda Facts," *Pandas International*, 2014, section "Reproduction," https://web.archive.org/web/20150924063658/http://www.pandasinternational.org/wptemp/education-2/panda-facts/.

16. David L. Hull, "The God of the Galápagos," *Nature* 352 (August 8, 1991): 485–486.

17. John Paul II, *Catechism of the Catholic Church* [1993], 2nd ed. (Washington DC: United States Catholic Conference, 2011), para. 400, https://www.usccb.org/sites/default/files/flipbooks/catechism/.

18. Daniel 10:13, 10:21, and 12:1; Jude 1:9; and Revelation 12:7.

19. C. S. Lewis, *The Problem of Pain* (New York: Simon and Schuster, 1996),119 and 122–123.

20. Hugh Ross, *Improbable Planet* (Grand Rapids, MI: Baker Books, 2016), 27.

21. Revelation 21:5.

22. Sagan published a book with the title *Pale Blue Dot: A Vision of the Human Future in Space* (New York: Random House, 1994).

23. Steven Weinberg, *The First Three Minutes* (New York: Bantam, 1977), 144.

24. Jonathan Wells, "Optimal Optics," *Salvo* 43 (Winter 2017), 45–47.

25. "Be Grateful for the Intelligent Design of Your Eyes," Discovery Science, *YouTube*, November 20, 2017, video, 9:19, https://www.youtube.com/watch?v=kboUBQnMP8w&t=5s.
26. Philip Gibbs, "Can a Human See a Single Photon?," *Physics FAQ*, 1996, https://math.ucr.edu/home/baez/physics/Quantum/see_a_photon.html.
27. See Casey Luskin, "The Recurrent Laryngeal Nerve Does Not Refute Intelligent Design," *IDEA Center*, accessed July 25, 2024, http://www.ideacenter.org/contentmgr/showdetails.php/id/1507 and Wolf-Ekkehard Lönnig, "The Laryngeal Nerve of the Giraffe: Does it Prove Evolution?," *Wolf-Ekkehard Lönnig*, September 7, 2010, https://www.weloennig.de/LaryngealNerve.pdf.

6. Prehistoric Man

1. George Gaylord Simpson, *The Meaning of Evolution*, rev. ed. (New Haven: Yale University Press, 1967), 345.
2. Benedict XVI, "Homily of His Holiness Benedict XVI," St. Peter's Square, Vatican City, April 24, 2005, https://www.vatican.va/content/benedict-xvi/en/homilies/2005/documents/hf_ben-xvi_hom_20050424_inizio-pontificato.html.
3. American Museum of Natural History, "Review: Most Human Origins Stories Are Not Compatible with Known Fossils," *Phys.org*, May 6, 2021, https://phys.org/news/2021-05-human-stories-compatible-fossils.html. The research paper being discussed is Sergio Almécija et al., "Fossil Apes and Human Evolution," *Science* 373, no. 6542 (May 7, 2021), https://www.science.org/doi/10.1126/science.abb4363. See also Günter Bechly, "Scientists Conclude: Human Origins Research Is a Big Mess," *Evolution News & Science Today*, May 10, 2021, https://evolutionnews.org/2021/05/scientists-conclude-human-origins-research-is-a-big-mess/.
4. "*Homo neanderthalensis*," *Smithsonian National Museum of Natural History*, accessed September 21, 2020, https://humanorigins.si.edu/evidence/human-fossils/species/homo-neanderthalensis. It turns out that primitive human fossils had been found all the way back in 1829, but they were not identified as such until several years after the findings near Dusseldorf.
5. "List of Neanderthal Fossils," *Wikipedia*, July 28, 2204, https://en.wikipedia.org/wiki/List_of_Neanderthal_fossils. For the early date of 400,000 years ago for the appearance of Neanderthals, see Lisa Hendry, "Who Were the Neanderthals?," *Natural History Museum*, accessed September 2024, https://www.nhm.ac.uk/discover/who-were-the-neanderthals.html.
6. Adam P. Van Arsdale, "*Homo erectus*—A Bigger, Smarter, Faster Hominin Lineage," *Nature Education* 4, no. 1 (2013), https://www.nature.com/scitable/knowledge/library/homo-erectus-a-bigger-smarter-97879043/.
7. Katerina Douka et al., "Age Estimates for Hominin Fossils and the Onset of the Upper Palaeolithic at Denisova Cave," *Nature* 565 (January 31, 2019): 640–644, https://doi.org/10.1038/s41586-018-0870-z.
8. Simon Fisher, quoted in Matthew Warren, "Biggest Denisovan Fossil Yet Spills Ancient Human's Secrets," *Nature* 569 (May 1, 2019): 16–17, https://www.nature.com/articles/d41586-019-01395-0.

9. Colin Barras, "Baboon Bone Found in Famous Lucy Skeleton," *New Scientist* (April 10, 2015), https://www.newscientist.com/article/dn27325-baboon-bone-found-in-famous-lucy-skeleton/.
10. Ann Gibbons, "A New Kind of Ancestor: *Ardipithecus* Unveiled," *Science* 326 (October 2, 2009): 36–40. For further discussion of these and the other much-hyped fossils mentioned in this chapter and said to be ancestral to humans, see Ann Gauger, Douglas Axe, and Casey Luskin, *Science and Human Origins* (Seattle, WA: Discovery Institute Press, 2012). This chapter benefits significantly from that book.
11. Esteban E. Sarmiento, "Comment on the Paleobiology and Classification of *Ardipithecus ramidus*," *Science* 328, no. 1105b (May 28, 2010).
12. Thomas C. Prang et al., "*Ardipithecus* Hand Provides Evidence That Humans and Chimpanzees Evolved from an Ancestor with Suspensory Adaptations," *Science Advances* 7, no. 9 (February 24, 2021), https://www.science.org/doi/10.1126/sciadv.abf2474. See also Casey Luskin, "Study: Hands of 'Ardi' Indicate a Chimp-like Tree-Dweller and Knuckle-Walker," *Evolution News & Science Today*, February 26, 2021, https://evolutionnews.org/2021/02/study-hands-of-ardi-indicate-a-chimp-like-tree-dweller-and-knuckle-walker/.
13. Quoted in John Noble Wilford, "Scientists Challenge 'Breakthrough' on Fossil Skeleton," *New York Times*, May 27, 2010, https://www.nytimes.com/2010/05/28/science/28fossil.html.
14. Bernard Wood, "Hominid Revelations from Chad," *Nature* 418 (July 11, 2002): 133–135.
15. Bernard Wood and Mark Grabowski, "Macroevolution in and around the Hominin Clade," *Macroevolution: Explanation, Interpretation, and Evidence*, eds. Serrelli Emanuele and Nathalie Gontier (Heidelberg, Germany: Springer, 2015), 347–376.
16. Quoted in Luskin, "Human Origins and the Fossil Record," in *Science and Human Origins*, 45. Luskin does not give a public source for the information.
17. Roberto Macchiarelli et al., "Nature and Relationships of *Sahelanthropus tchadensis*," *Journal of Human Evolution* 149 (December 2020), https://doi.org/10.1016/j.jhevol.2020.102898.
18. Frans B. M. de Waal, "Apes from Venus: Bonobos and Human Social Evolution," in *Tree of Origin: What Primate Behavior Can Tell Us about Human Social Evolution*, ed. Frans B. M. de Waal (Cambridge: Harvard University Press, 2001), 68.
19. Mark Davis, "Into the Fray: The Producer's Story," *PBS NOVA Online*, February 2002, http://www.pbs.org/wgbh/nova/neanderthals/producer.html.
20. Seth Borenstein, "Fossil Discovery Fills Gap in Human Evolution," MSNBC, April 12, 2006, https://www.nbcnews.com/id/wbna12286206.
21. Ernst Mayr, *What Makes Biology Unique?* (New York: Cambridge University Press, 2004), 198.
22. See Gauger, "Science and Human Origins" in *Science and Human Origins*, 21–22.
23. David Reich, *Who We Are and How We Got Here* (New York: Pantheon, 2018), 96.
24. See, for example, "Genetics," *Smithsonian Museum of Natural History*, accessed September 21, 2020, https://humanorigins.si.edu/evidence/genetics.

25. Casey Luskin, "Evolutionary Models of Palaeoanthropology, Genetics, and Psychology Fail to Account for Human Origins: A Review," in *Science and Faith in Dialogue*, eds. Frikkie Van Niekerk and Nico Vorster (Cape Town, South Africa: Aosis, 2022), 243–282.
26. Yuxin Fan et al., "Genomic Structure and Evolution of the Ancestral Chromosome Fusion Site in 2q13-2q14.1 and Paralogous Regions on Other Human Chromosomes," *Genome Research* 12 (2002): 1651–1662. See also Ann K. Gauger, Ola Hössjer, and Colin R. Reeves, "Evidence for Human Uniqueness," in *Theistic Evolution*, ed. J. P. Moreland et al., (Wheaton, IL: Crossway, 2017). See also Luskin, "Francis Collins, Junk DNA, and Chromosomal Fusion," in *Science and Human Origins*, 86–104.
27. Luskin, "Francis Collins, Junk DNA, and Chromosomal Fusion."
28. Luskin, "Francis Collins," 91; Luskin, "Lessons Learned (and Not Learned) from the Evangelical Debate over Adam and Eve," *Christian Research Journal* 45 (2022); Luskin, "Evolutionary Models of Palaeoanthropology"; Luskin, "Comparing Contemporary Evangelical Models Regarding Human Origins," *Religions* 14, no. 6 (2023): 748, https://doi.org/10.3390/rel14060748.
29. Reich, *Who We Are*, 31–32.
30. Ewen Callaway, "Genetic Adam and Eve Did Not Live Too Far Apart in Time," *Nature* (August 6, 2013), https://www.nature.com/news/genetic-adam-and-eve-did-not-live-too-far-apart-in-time-1.13478.
31. Callaway, "Genetic Adam and Eve."
32. Gauger, "The Science of Adam and Eve," in *Science and Human Origins*, 103–122.
33. Dennis R. Venema and Scot McKnight, *Adam and the Genome: Reading Scripture after Genetic Science* (Grand Rapids, MI: Brazos Press, 2017).
34. Casey Luskin, "Lessons from the Evangelical Debate about Adam and Eve," *Evolution News & Science Today*, November 15, 2021, https://evolutionnews.org/2021/11/lessons-from-the-evangelical-debate-about-adam-and-eve/. The first remark by Haarsma appears in the editorial note in Thomas H. McCall, "Will The Real Adam Please Stand Up? The Surprising Theology Of Universal Ancestry," *BioLogos*, March 23, 2020, https://biologos.org/series/book-review-the-genealogical-adam-and-eve/articles/will-the-real-adam-please-stand-up-the-surprising-theology-of-universal-ancestry. The second can be found at Deborah Haarsma, "Truth-Seeking in Science," *BioLogos*, January 10, 2020, https://biologos.org/articles/truth-seeking-in-science.
35. Reich, *Who We Are*, 77.
36. Reich, *Who We Are*, 49.
37. Matthew Warren, "Diverse Genome Study Upends Understanding of How Language Evolved," *Nature News* (August 2, 2018), https://www.nature.com/articles/d41586-018-05859-7.
38. Reich, *Who We Are*, 36.
39. Luskin, "Human Origins and the Fossil Record," in *Science and Human Origins*, 71.
40. Sonia Harmand et al., "3.3-Million-Year-Old Stone Tools from Lomekwi 3, West Turkana, Kenya," *Nature* 521 (May 21, 2015): 310–315, https://www.nature.com/articles/nature14464.

41. Thomas W. Plummer et al., "Expanded Geographic Distribution and Dietary Strategies of the Earliest Oldowan Hominins and *Paranthropus*," *Science* 379 (February 10, 2023): 561–566.
42. Margherita Mussi et al., "Early *Homo erectus* Lived at High Altitudes and Produced Both Oldowan and Acheulean Tools," *Science* 382, no. 6671 (October 12, 2023): 713–718, https://www.science.org/doi/10.1126/science.add9115.
43. James Blinkhorn et al., "Constraining the Chronology and Ecology of Late Acheulean and Middle Palaeolithic Occupations at the Margins of the Monsoon," *Scientific Reports* 11 (2021), https://www.nature.com/articles/s41598-021-98897-7.
44. L. Barham et al., "Evidence for the Earliest Structural Use of Wood at Least 476,000 Years Ago," *Nature* 622 (2023), 107–111, https://doi.org/10.1038/s41586-023-06557-9.
45. Simon A. Parfitt and Silvia M. Bello, "Bone Tools, Carnivore Chewing and Heavy Percussion: Assessing Conflicting Interpretations of Lower and Upper Palaeolithic Bone Assemblages," *Royal Society Open Science* 11, no. 1 (January 2024), http://doi.org/10.1098/rsos.231163.
46. "Middle Stone Age Tools," *Smithsonian Museum of Natural History*, accessed October 1, 2020, https://humanorigins.si.edu/evidence/behavior/stone-tools/middle-stone-age-tools.
47. Matt Kaplan, "Million-Year-Old Ash Hints at Origins of Cooking," *Nature News*, April 2, 2012, https://www.nature.com/articles/nature.2012.10372.
48. C. K. Brain and A. Sillent, "Evidence from the Swartkrans Cave for the Earliest Use of Fire," *Nature* 366 (December 1, 1988): 464–466.
49. Xing Gao et al., "Evidence of Hominin Use and Maintenance of Fire at Zhoukoudian," *Current Anthropology*, 58 (August 2017): S267–277.
50. Wil Roebroeks and Paola Villa, "On the Earliest Evidence for Habitual Use of Fire in Europe," *PNAS* 108, no. 13 (March 14, 2011): 5209–5214; Francesco Berna and Paul Goldberg, "Assessing Paleolithic Pyrotechnology and Associated Hominin Behavior in Israel," *Israeli Journal of Earth Sciences* 56 (2007): 107–121; Francesco Berna et al., "Microstratigraphic Evidence of In Situ Fire in the Acheulean Strata of Wonderwerk Cave, Northern Cape Province, South Africa," *PNAS* 109, no. 20 (April 2, 2012): E1215-E1220.
51. This whole section borrows heavily from a review article by J. A. J. Gowlett, "The Discovery of Fire by Humans: A Long and Convoluted Process," *Philosophical Transactions of the Royal Society B* 371 (June 5, 2016), https://doi.org/10.1098/rstb.2015.0164.
52. Lawrence E. Murr, "A Brief History of Metals," in *Handbook of Materials Structures, Properties, Processing and Performance* (Switzerland: Springer, 2015), 3–9.
53. Miljana Radivojević et al., "On the Origins of Extractive Metallurgy: New Evidence from Europe," *Journal of Archaeological Science* 37, no. 11 (November 2010): 2775–3787.
54. Murr, "A Brief History of Metals." See also Mehmet Özdoğan, "The Making of the Early Bronze Age in Anatolia," *Old World: Journal of Ancient Africa and Eurasia* 3, no. 1 (October 2023): 1–58.

55. Alan W. Pense, "Iron through the Ages," *Materials Characterization* 45, no. 4–5 (October–November 2000): 353–363.
56. Bruce Bower, "Glassmaking May Have Begun in Egypt, Not Mesopotamia," *Science News* (November 22, 2016), https://www.sciencenews.org/article/glassmaking-may-have-begun-egypt-not-mesopotamia.
57. "Currently scholars believe that glass was discovered either as a byproduct of metallurgy or from an evolutionary sequence in the development of ceramic materials," according to "Origins of Glass: Myth and Known History," in Seth C. Rasmussen, *How Glass Changed the World: The History and Chemistry of Glass from Antiquity to the 13th Century* (London: Springer-Verlag, 2012), 15.
58. A. W. G. Pike et al., "U-Series Dating of Paleolithic Art in 11 Caves in Spain," *Science* 336 (June 15, 2012): 1409–1413.
59. William K. Jones and Lee F. Elliott, "Art in European Caves," in *Encyclopedia of Caves*, ed. William B. White et al. (London: Academic Press, 2019), 71–75, https://doi.org/10.1016/B978-0-12-814124-3.00010-8.
60. Dirk L. Hoffmann et al., "U-Th Dating of Carbonate Crusts Reveals Neandertal Origin of Iberian Cave Art," *Science* 359 (February 23, 2018): 912–915, https://www.science.org/doi/10.1126/science.aap7778.
61. Dirk L. Hoffmann et al., "Symbolic Use of Marine Shells and Mineral Pigments by Iberian Neandertals 115,000 Years Ago," *Science Advances* 4, no. 2 (February 22, 2018), https://www.science.org/doi/10.1126/sciadv.aar5255.
62. Hoffmann et al., "U-Th Dating."
63. Jacques Jaubert et al., "Early Neanderthal Constructions Deep in Bruniquel Cave in Southwestern France," *Nature* 534 (May 25, 2016): 111–114, https://doi.org/10.1038/nature18291.
64. Jaubert et al., "Early Neanderthal Constructions."
65. Rebecca Wragg Sykes, *Kindred: Neanderthal Life, Love, Death and Art* (London: Bloomsbury, 2020), 246.
66. Adam Brumm et al., "Oldest Cave Art Found in Sulawesi," *Science Advances* 7, no. 3 (January 13, 2021), https://www.science.org/doi/10.1126/sciadv.abd4648.
67. Bruno David et al., "A 28,000 Year Old Excavated Painted Rock from Nawarla Gabarnmang, Northern Australia," *Journal of Archaeological Science* 40, no. 5 (May 2013): 2493–2501, https://doi.org/10.1016/j.jas.2012.08.015.
68. Lorna Till, *Theory and Practice in the Bioarchaeology of Care* (New York: Springer, 2015), 219.
69. Christopher S. Henshilwood et al., "An Abstract Drawing from the 73,000-Year-Old Levels at Blombos Cave, South Africa," *Nature* 562 (October 2018): 115–117; Michael Balter, "On the Origin of Art and Symbolism," *Science* 323 (February 6, 2009): 709–711.
70. "It Must Be a Woman," University of Tübingen, July 22, 2015, https://web.archive.org/web/20161011145105/https://www.uni-tuebingen.de/en/news/press-releases/newsfullview-pressemitteilungen/article/es-muss-eigentlich-eine-frau-sein.html.
71. Paul Rincon, "'Oldest Sculpture' Found in Morocco," *BBC Science*, May 23, 2003, http://news.bbc.co.uk/1/hi/sci/tech/3047383.stm.

72. Hansjörg Hemminger, *Evolutionary Processes in the Natural History of Religion* (Switzerland: Springer, 2021), 108–109; Rex Dalton, "Lion Man Takes Pride of Place as Oldest Statue," *Nature* 425 (September 4, 2003): 7.
73. For a discussion see Francesco D'Errico et al., "A Middle Palaeolithic Origin of Music? Using Cave-Bear Bone Accumulations to Assess the Divje Babe I Bone 'Flute,'" *Antiquity* 72, no. 275 (March 1998): 65–79; See also "Hear the World's Oldest Instrument, the 'Neanderthal Flute,' Dating Back Over 43,000 Years," *Open Culture*, February 10, 2015, http://www.openculture.com/2015/02/hear-the-worlds-oldest-instrument-the-neanderthal-flute.html.
74. "Earliest Music Instruments Found," *BBC News*, May 25, 2012, https://www.bbc.com/news/science-environment-18196349; Nicholas J. Conard et al., "New Flutes Document the Earliest Musical Tradition in Southwestern Germany," *Nature* 460 (2009): 737–740; Thomas Higham et al., "Testing Models for the Beginnings of the Aurignacian and the Advent of Figurative Art and Music: The Radiocarbon Chronology of Geißenklösterle," *Journal of Human Evolution* 62, no. 6 (June 2012): 664–676.
75. "Prehistoric Clothing," *Encyclopedia of Fashion*, accessed October 2020, http://www.fashionencyclopedia.com/fashion_costume_culture/The-Ancient-World-Prehistoric/Prehistoric-Clothing.html.
76. "UF Study of Lice DNA Shows Humans First Wore Clothes 170,000 Years Ago," *University of Florida News*, January 6, 2011, https://news.ufl.edu/archive/2011/01/uf-study-of-lice-dna-shows-humans-first-wore-clothes-170000-years-ago.html; Melissa A. Toups et al, "Origin of Clothing Lice Indicates Early Clothing Use by Anatomically Modern Humans in Africa," *Molecular Biology and Evolution* 28, no.1 (2011): 29–32.
77. David W. Frayer et al., "Krapina and the Case for Neandertal Symbolic Behavior," *Current Anthropology* 61, no. 6 (December 2020): 713–731; Herbert Ullrich, "Cannibalistic Rites within Mortuary Practices from the Paleolithic to Middle Ages in Europe," *Anthropologie* 43 (2005): 249–261.
78. Emma Pomeroy et al., "New Neanderthal Remains Associated with the 'Flower Burial' at Shanidar Cave," *Antiquity* 94, no. 373 (April 2020), 11–26.
79. Daniella E. Bar-Yosef Mayer, Bernard Vandermeersch, and Ofer Bar-Ysoef, "Shells and Ochre in Middle Paleolithic Qafzeh Cave, Israel: Indications for Modern Behavior," *Journal of Human Evolution* 56 (2009): 307–314.
80. James M. Bowler et al., "New Ages for Human Occupation and Climatic Change at Lake Mungo, Australia," *Nature* 421 (2003): 837–840; M. Barbetti and H. Allen, "Prehistoric Man at Lake Mungo, Australia, by 32,000 Years BP," *Nature* 240 (November 3, 1972): 46–48.
81. Maria Martinón-Torres et al., "Earliest Known Human Burial in Africa," *Nature* 593 (May 5, 2021): 95–100.
82. Michael Balter, "First Jewelry? Old Shell Beads Suggest Early Use of Symbols," *Science* 312, no. 5781 (June 23, 2006): 1731; Marian Vanhaereny et al., "Middle Paleolithic Shell Beads in Israel and Algeria," *Science* 312, no. 5781 (June 23, 2006): 1785–1788.
83. Hoffmann et al., "Symbolic Use of Marine Shells."

84. S. McBrearty and A. S. Brooks, "The Revolution That Wasn't: A New Interpretation of the Origin of Modern Human Behavior," *Journal of Human Evolution* 39 (2000): 453–563.
85. Martin Sikora et al., "Ancient Genomes Show Social and Reproductive Behavior of Early Upper Paleolithic Foragers," *Science* 358 (2017): 659–662; Maria Dobrovolskaya et al., "Direct Radiocarbon Dates for the Mid Upper Paleolithic (Eastern Gravettian) Burials from Sunghir, Russia," *Bulletins et Mémoires de la Société d'anthropologie de Paris* 24 (2012): 96–102.
86. See Merritt Ruhlen, *The Origin of Language: Tracing the Evolution of the Mother Tongue* (New York: Wiley, 1994), 150.
87. See Reich, *Who We Are*, 197–198.
88. For an overview and discussion of recent mainstream scientific literature calling into question *Homo habilis*'s place in the genus *homo*, see Günter Bechly, "Fossil Friday: New Research Questions the Human Nature of *Homo habilis*," *Evolution News & Science Today*, January 19, 2024, https://evolutionnews.org/2024/01/fossil-friday-new-research-questions-the-human-nature-of-homo-habilis/. See also Casey Luskin, "Evolutionary Models of Palaeoanthropology, Genetics and Psychology Fail to Account for Human Origins: A Review," in *Science and Faith in Dialogue*, eds. Frikkie Van Niekerk and Nico Vorster (Cape Town, South Africa: Aosis, 2022). For a discussion of *Homo naledi* and why it probably does not belong in *Homo*, see Casey Luskin, "Hominid Hype and *Homo naledi*: A Unique 'Species' of Unclear Evolutionary Importance," *Evolution News & Science Today*, September 20, 2015, https://evolutionnews.org/2015/09/hominid_hype_an/. For a discussion of how even mainstream scientists don't agree on how *Homo naledi* is related to other species, see "What Does It Mean to Be Human?," *Smithsonian Museum of Natural History*, January 3, 2024, https://humanorigins.si.edu/evidence/human-fossils/species/homo-naledi.
89. Piotr Lenartowicz, *Ludy czy małpoludy? [People or Manapes?*, English summary of Polish text] (Krakow, Poland: Jesuit University of Philosophy and Education Ignatianum, 2010), 260, https://lenartowicz.jezuici.pl/files/2011/09/54-SUMMARY1.pdf. The work initially circulated as a PDF, of which the URL here is a portion. In a later bound version of the book, the quotation appears on p. 383.
90. Sermonti, *Why Is a Fly not a Horse?*, 71–72.
91. "Facts at a Glance," *Pontifical Academy of Sciences*, accessed September 12, 2024, https://www.pas.va/en/about.html.
92. Vivien Gornitz, "Sea Level Rise, After the Ice Melted and Today," *NASA*, January 2007, https://www.giss.nasa.gov/research/briefs/archive/2007_gornitz_09/.
93. The burgeoning field of submarine archaeology has discovered much information about how humans lived during periods when sea levels were much lower. See Isaac Ogloblin Ramirez, Ehud Galili, and Ruth Shahack-Gross, "Locating Submerged Prehistoric Settlements: A New Underwater Survey Method Using Water-Jet Coring and Micro-geoarchaeological Techniques," *Journal of Archaeological Science* 135 (November 2021), https://www.sciencedirect.com/science/article/abs/pii/S0305440321001503. However, these discoveries tend to be from more recent times. Given the challenges of preserving artifacts under water for hundreds of thousands, or even millions of years, this is not surprising.

7. Man: The Image of God

1. John Paul II, "Message to the Pontifical Academy of Science on Evolution," October 22, 1996, section 5. Available at EWTN, https://www.ewtn.com/catholicism/library/message-to-the-pontifical-academy-of-science-on-evolution-8825.
2. On the psychological analogy see Books 9–15 of Augustine, *De Trinitate (On the Trinity)* [c. 417], trans. Arthur West Haddan, in *Nicene and Post-Nicene Fathers, First Series*, vol. 3, ed. Philip Schaff (Buffalo, NY: Christian Literature Publishing, 1887), https://www.newadvent.org/fathers/1301.htm.
3. Thomas Aquinas, *Summa Theologiae* [1274] I, question 32, article 1, objection 2 reply. See *The Summa Theologiæ of St. Thomas Aquinas*, trans. Fathers of the English Dominican Province [1920], https://www.newadvent.org/summa/1032.htm.
4. Charles Darwin to William Graham, July 3, 1881, *Darwin Correspondence Project*, Letter no. 13230, University of Cambridge, https://www.darwinproject.ac.uk/letter/?docId=letters/DCP-LETT-13230.xml.
5. Benjamin Wiker, *The Darwin Myth: The Life and Lies of Charles Darwin*, (Washington, DC: Regnery Publishing, 2009), 107–109.
6. Charles Darwin to J. D. Hooker, February 24–25, 1863, *Darwin Correspondence Project*, Letter no. 4009, University of Cambridge, https://www.darwinproject.ac.uk/letter/?docId=letters/DCP-LETT-4009.xml.
7. Alfred Russel Wallace, "Sir Charles Lyell on Geological Climates and the Origin of Species," *Quarterly Review* (April 1869): 391.
8. Evelyn Fox Keller, *Making Sense of Life: Explaining Biological Development with Models, Metaphors, and Machines* (Cambridge, MA: Harvard University Press, 2002), 295–296.
9. John Paul II, *Catechism of the Catholic Church* [1993], 2nd ed. (Washington DC: United States Catholic Conference, 2011), para. 31, https://www.usccb.org/sites/default/files/flipbooks/catechism/.
10. Henry George Liddell and Robert Scott, *Greek-English Lexicon* [1843] (Oxford, UK: Oxford University Press, 1940).
11. Noam Chomsky, "Things No Amount of Learning Can Teach," interview by John Gliedman, *Omni* 6, no. 11, November 1983, https://chomsky.info/198311.
12. Noam Chomsky and James McGilvray, *The Science of Language: Interviews with James McGilvray* (New York: Cambridge University Press, 2012), 13.
13. See Margot Adler, "The Chimp That Learned Sign Language," NPR, May 28, 2008, https://www.npr.org/2008/05/28/90516132/the-chimp-that-learned-sign-language.
14. Michael Denton, *Evolution: A Theory in Crisis* (Bethesda, MD: Adler & Adler, 1986), 170.
15. Norman Doidge, *The Brain That Changes Itself* (New York: Viking, 2007), 258.
16. See William Reville, "Remarkable Story of Maths Genius Who Had Almost No Brain," *Irish Times*, November 9, 2006, https://www.irishtimes.com/news/remarkable-story-of-maths-genius-who-had-almost-no-brain-1.1026845.
17. L. von Melchner, S. L.Pallas, and M. Sur, "Visual Behaviour Mediated by Retinal Projections Directed to the Auditory Pathway," *Nature* 404, no. 6780 (April 20,

2000), 871–876, https://sci-hub.se/10.1038/35009102. Cited by Denton, *Evolution: A Theory in Crisis*, 171.

18. Chomsky and McGivray, *The Science of Language*, 13.
19. Rodney Stark, *Discovering God: The Origin of the Great Religions and the Evolution of Belief* (Toronto: Harper Collins, 2007), 61–62.
20. See the section on de Bonald in Etienne Gilson, Thomas Langan, and Armand Maurer, *Recent Philosophy, vol 1: From Hegel to Sartre* (Providence, RI: Cluny, 2023), esp. 236.
21. David Premack, "Gavagai! Or the Future of the Animal Language Controversy," *Cognition* 19, no. 3 (1985): 281–282, https://doi.org/10.1016/0010-0277(85)90036-8.
22. Chomsky and McGivray, *The Science of Language*, 49.
23. Marc Hauser et al., "The Mystery of Language Evolution," *Frontiers in Psychology*, 7 May 2014.
24. "Heritage Minutes: Wilder Penfield," Historica Canada, *YouTube*, March 2, 2016, video, 1:01, https://www.youtube.com/watch?v=pUOG2g4hj8s.
25. Wilder Penfield, *The Mystery of the Mind: A Critical Study of Consciousness and the Human Brain* (Princeton, NJ: Princeton University Press, 1975), 77–78.
26. Penfield, *The Mystery of the Mind*, 55.
27. Michael Egnor, "Pioneering Neuroscientist Wilder Penfield: Why Don't We Have Intellectual Seizures," *Evolution News & Science Today*, April 21, 2016, https://evolutionnews.org/2016/04/wilder_penfield/.
28. John Eccles and Daniel N. Robinson, *The Wonder of Being Human: Our Brain and Our Mind* (New York: Free Press, 1984), 36.
29. Matthew Cobb, "Why Your Brain Is Not a Computer," *Guardian*, February 27, 2020, https://www.theguardian.com/science/2020/feb/27/why-your-brain-is-not-a-computer-neuroscience-neural-networks-consciousness. This article is an edited excerpt from Cobb's book *The Idea of the Brain* (New York: Basic Books, 2020). It is discussed at Michael Egnor, "The Mind Is the Opposite of a Computer," *Mind Matters*, March 5, 2020, https://mindmatters.ai/2020/03/the-mind-is-the-opposite-of-a-computer/.
30. Steve Taylor, "Near Death Experiences and DMT," *Psychology Today*, October 12, 2018, https://www.psychologytoday.com/us/blog/out-the-darkness/201810/near-death-experiences-and-dmt.
31. Christof Koch, "What Near-Death Experiences Reveal about the Brain," *Scientific American*, June 1, 2020, https://www.scientificamerican.com/article/what-near-death-experiences-reveal-about-the-brain/.
32. Gary Habermas provides a brief review of the recent research on negative near-death experiences, and their possible under-reporting: "Evidential Near-Death Experiences," in *Minding the Brain: Models of the Mind, Information, and Empirical Science* (Seattle, Washington: Discovery Institute Press, 2023), 352 n92.
33. See Kenneth Ring and Sharon Cooper, "Near-Death and Out-of-Body Experiences in the Blind: A Study of Apparent Eyeless Vision," *Journal of Near-Death Studies* 16, no. 2 (Winter 1997): 101–147.

34. See Pim van Lommel, *Consciousness Beyond Life: The Science of the Near-Death Experience* (New York: Harper Collins, 2010), 71–72.
35. Neal Grossman, "Who's Afraid of Life After Death?," *Journal of Near-Death Studies* 21, no. 1 (Fall 2002): 8. Cited in Mario Beauregard and Denyse O'Leary, *The Spiritual Brain: A Neuroscientist's Case for the Existence of the Soul* (New York: Harper Collins, 2007), 166.
36. P. Ringger, "Die Mystik im Irrsinn," *Neue Wissenschaft 8* (1958): 217–220. Cited in Michael Nahm and Bruce Greyson, "The Death of Anna Katharina Ehmer: A Case Study in Terminal Lucidity," *OMEGA—Journal of Death and Dying* 68, no. 1 (January 2013): 77–87, https://www.researchgate.net/publication/260250637_The_Death_of_Anna_Katharina_Ehmer_A_Case_Study_in_Terminal_Lucidity.
37. Nahm and Greyson, "The Death of Anna Katharina Ehmer: A Case Study in Terminal Lucidity," 77–87.
38. Michael Nahm et al., "Terminal Lucidity: A Review and a Case Collection," *Archives of Gerontology and Geriatrics* 55, no. 1 (July–August 2012): 138–142, https://www.sciencedirect.com/science/article/abs/pii/S0167494311001865?via%3Dihub. Text available at University of Virginia School of Medicine, Division of Perceptual Studies, https://med.virginia.edu/perceptual-studies/wp-content/uploads/sites/360/2016/12/OTH25terminal-lucidity-AGG.pdf.
39. J. B. Moseley et al., "Arthroscopic Treatment of Osteoarthritis of the Knee: A Prospective, Randomized, Placebo-Controlled Trial. Results of a Pilot Study," *American Journal of Sports Medicine* 24, no. 1 (January–February 1996): 28–34.
40. Beauregard and O'Leary, *The Spiritual Brain*, 145.
41. "Artificial General Intelligence," *Wikipedia*, last modified October 2, 2024, https://en.wikipedia.org/wiki/Artificial_general_intelligence#Philosophical_perspective.
42. George D. Montañez, "Detecting Intelligence: The Turing Test and Other Design Detection Methodologies," *Proceedings of the Eighth International Conference on Agents and Artificial Intelligence* (January 2016): 517–523, https://www.scitepress.org/papers/2016/58237/58237.pdf.
43. See George Montañez, "Bingecast: George Montañez on Intelligence and the Turing Test," interview by Robert J. Marks, *Mind Matters*, June 25, 2020, podcast, audio, 18:57, https://mindmatters.ai/podcast/ep88/.
44. George Dvorsky, "Why the Turing Test Is Bullsh*t," *Gizmodo*, June 9, 2014, https://gizmodo.com/why-the-turing-test-is-bullshit-1588051412.
45. Eric Larson, *The Myth of Artificial Intelligence* (Cambridge, MA: Belknap Press, 2021), 192–193.
46. Anh Nguyen, Jason Yoskinki, and Jeff Clune, "Deep Neural Networks are Easily Fooled: High Confidence Predictions for Unrecognizable Images," *Proceedings of the Institute of Electrical and Electronics Engineers Conference on Computer Vision and Pattern Recognition* (2015): 427–436, https://www.computer.org/csdl/proceedings/cvpr/2015/12OmNBkfRhw. Text available at *Arxiv*, https://arxiv.org/abs/1412.1897.

47. Larson, *The Myth of Artificial Intelligence*, 53.
48. Larson, *The Myth of Artificial Intelligence*, 199–200.
49. David Cole, "The Chinese Room Argument," *Stanford Encyclopedia of Philosophy* (Summer 2023), https://plato.stanford.edu/entries/chinese-room/.
50. See Stephen M. Barr, "The Atheism of the Gaps," *First Things*, November 1995.
51. Hao Wang, *A Logical Journey: From Gödel to Philosophy* (Cambridge, MA: MIT Press, 1996), section 6.2.11.
52. Gregory Chirikjian, "Help Wanted: For the Cognitive Era," JHUENGINEERING, Summer 2017, https://engineering.jhu.edu/magazine/2017/05/help-wanted-for-the-cognitive-era/.
53. Satya Nadella with Grew Shaw and Jill Tracie Nichols, *Hit Refresh: The Quest to Rediscover Microsoft's Soul and Imagine a Better Future for Everyone* (New York: Harper Business, 2017), 207.
54. Augustine, De *Trinitate* (*On the Trinity*), Books 9–15.
55. Aquinas, *Summa Theologiae* I, question 94, article 3, https://www.newadvent.org/summa/1094.htm.

8. Anti-Theist Darwin and His Useful Instruments

1. Giuseppe Sermonti, *Why Is a Fly Not a Horse?* (Seattle, WA: Discovery Institute, 2005), 11.
2. Bonaventure, *Bonaventure: The Soul's Journey into God / The Tree of Life / The Life of St. Francis* [1217–1274], trans. Ewert Cousins (New York: Paulist Press, 1978), 67.
3. Daniel Dennett, *Darwin's Dangerous Idea: Evolution and the Meanings of Life* (New York: Touchstone, 1995), 18.
4. Charles Darwin to Asa Gray, May 22, 1860, *Darwin Correspondence Project*, Letter no. 2814, University of Cambridge, https://www.darwinproject.ac.uk/letter/?docId=letters/DCP-LETT-2814.xml.
5. Benjamin Wiker, *The Darwin Myth: The Life and Lies of Charles Darwin* (Washington, DC: Regnery Publishing, 2009), 1.
6. Desmond King-Hele, *Erasmus Darwin: A Life of Unequalled Achievement* (London: DLM, 1999), 301–302.
7. Wiker, *The Darwin Myth*, 17.
8. Adrian Desmond and James Moore, *Darwin: The Life of a Tormented Evolutionist* (New York: Norton, 1994), 34.
9. Michael Flannery, *Alfred Russel Wallace, Intelligent Evolution: How Alfred Russel Wallace's World of Life Challenged Darwinism*, (Georgetown, KY: Erasmus Press, 2020), 28. For a fuller discussion, see Howard E. Gruber, *Darwin on Man: A Psychological Study of Scientific Creativity* (New York: E. P. Dutton, 1974), 39 and 479.
10. R. B. Freeman, "Discourse to the Plinian Society: An Introduction," in Freeman, *The Works of Charles Darwin* (Folkestone, England: Dawson, 1977), https://darwin-online.org.uk/EditorialIntroductions/Freeman_DiscoursetothePlinian.html.

11. Charles Darwin, *Autobiography* [1887] ed. Nora Barlow (London: Collins, 1958), https://darwin-online.org.uk/content/frameset?itemID=F1497&viewtype=text&pageseq=1.
12. See Wiker, *The Darwin Myth*, 11. Also see Patricia Fara, "Myth 3—That Charles Darwin Was Not Directly Influenced by the Evolutionary Views of His Grandfather Erasmus," in *Darwin Mythology: Debunking Myths, Correcting Falsehoods*, ed. Kostas Kampourakis (New York: Cambridge University Press, 2024), https://www.cambridge.org/core/books/abs/darwin-mythology/that-charles-darwin-was-not-directly-influenced-by-the-evolutionary-views-of-his-grandfather-erasmus/69FF2F9AF9985F82EB8C3630330365B5#.
13. Desmond and Moore, *Darwin: The Life of a Tormented Evolutionist*, 656–658.
14. Charles Darwin, *Notebook C* [February–July 1838] available at *Darwin Online*, 166, https://darwin-online.org.uk/content/frameset?itemID=CUL-DAR122.-&viewtype=side&pageseq=1.
15. Flannery, *Intelligent Evolution*, 31.
16. Wiker, *The Darwin Myth*, 166.
17. Emma Darwin to Charles Darwin, c. February 1939, *Darwin Correspondence Project*, Letter no. 471, University of Cambridge, https://www.darwinproject.ac.uk/letter/?docId=letters/DCP-LETT-471.xml. Darwin's handwritten comment can be seen on the last page of this letter, reproduced at https://darwin-online.org.uk/content/frameset?itemID=CUL-DAR210.8.14&viewtype=image&pageseq=1.
18. Charles Darwin to William Graham, July 3, 1881, *Darwin Correspondence Project*, Letter no. 13230, University of Cambridge, https://www.darwinproject.ac.uk/letter/?docId=letters/DCP-LETT-13230.xml.
19. Pierre Teilhard de Chardin, *The Heart of Matter*, trans. René Hague (Harcourt: New York, 1979), 25.
20. Bruno de Solages, "Christianity and Evolution," *Cross Currents* (Summer 1951), quoted by John C. Greene, *Darwin and the Modern World View* (New York: New American Library, 1963), 77–78.
21. Raymond J. Nogar, "The Paradox of the Phenomenon," *Dominicana* 45 (Fall 1960): 248–249, quoted in Greene, *Darwin and the Modern World View*, 78. It should be noted the Fr. Nogar was not opposed to evolution itself, but to Teilhard's version.
22. P. B. Medawar, "Critical Notice," *Mind* 70, no. 277 (January 1961): 99.
23. Alasdair MacIntyre, *Guardian*, December 11, 1959, 7. Quoted in *Book Review Digest 1960* (New York: H. W. Wilson, 1961), 1324.
24. See Michael Chaberek, *Catholicism and Evolution: A History from Darwin to Pope Francis* (Kettering, OH: Angelico Press, 2015), 198.
25. Jonathan Webb, "Piltdown Review Points Decisive Finger at Forger Dawson," *BBC News*, August 10, 2016, https://www.bbc.com/news/science-environment-37021144.
26. Thomas O'Toole, "Piltdown Hoax Said to Involve Jesuit Scholar," *Washington Post*, July 15, 1980, https://www.washingtonpost.com/archive/politics/1980/07/16/piltdown-hoax-said-to-involve-jesuit-scholar/badaebe9-20b5-46c6-9d03-afe85d5d3da2/.

27. Chaberek, *Catholicism and Evolution*, 201.
28. Chaberek, *Catholicism and Evolution*, 209, note 29. See also John W. Flanagan, "A Periscope on Teilhard de Chardin," 3, https://traditioninaction.org/Questions/WebSources/B_303_Periscope%20on%20Teilhard.pdf.
29. "Monitum on the Writings of Fr. Teilhard de Chardin, SJ," June 30, 1962, available at *EWTN*, https://www.ewtn.com/catholicism/library/monitum-on-the-writings-of-fr-teilhard-de-chardin-sj-2144.
30. Cardinal Casaroli to Archbishop Paul Poupart, Rector of the Catholic Insitute of Paris, May 12, 1981, published in *L'Osservatore Romano*, June 10, 1981, available at https://www.traditioninaction.org/ProgressivistDoc/A_020_CasaroliTeilhard.htm.
31. "Teilhard de Chardin, Communique of Press Office of the Holy See," July 11, 1981, available at *EWTN*, https://www.ewtn.com/catholicism/library/teilhard-de-chardin-2595.
32. Benedict XVI, July 24, 2009. See Jim Manney, "Papal Praise for Teilhard," *Ignatian Spirituality*, July 30, 2009, https://www.ignatianspirituality.com/papal-praise-for-teilhard/.
33. Joseph Ratzinger, *Introduction to Christianity* (San Francisco, CA: Ignatius Press, 1990), 239. The book was first published in German in 1968 and in English in 1969.
34. C. S. Lewis to Fr Frederick Joseph Adelmann, September 21, 1960, in *The Collected Letters of C. S. Lewis, Volume 3: Narnia, Cambridge, and Joy, 1950–1963* (New York: HarperOne, 2007), 1186; C. S. Lewis to Bernard Acworth, March 5, 1960, *Collected Letters Vol. 3*, 1137.
35. See John L. Allen Jr., "Pope Cites Teilhardian Vision of the Cosmos as a 'Living Host,'" *National Catholic Reporter*, July 28, 2009, https://www.ncronline.org/news/pope-cites-teilhardian-vision-cosmos-living-host.
36. Pierre Teilhard de Chardin, *Phenomenon of Man* (New York: Harper Perennial, 1955), 311, 313.
37. In Chaberek, *Catholicism and Evolution*, 205.
38. Étienne Gilson to de Lubac, April 7, 1967, in *Letters of Étienne Gilson to Henri de Lubac: Annotated by Father de Lubac*, trans. Mary Emily Hamilton (San Francisco: Ignatius, 1988), 136–137.
39. *Gilson to de Lubac*, April 7, 1967, 134–135.
40. See Étienne Gilson, "Le Cas de Teilhard de Chardin," in *Les Tribulations de Sophie* (Paris: Vrin, 1967), 73–99, esp. 73–74.
41. *Gilson to de Lubac*, April 7, 1967, 59–60.
42. *Gilson to de Lubac*, April 7, 1967, 60–61.
43. *Gilson to de Lubac*, April 7, 1967, 61.
44. "Theodosius Dobzhansky," *Wikipedia*, last modified September 13, 2024, https://en.wikipedia.org/wiki/Theodosius_Dobzhansky.
45. Theodosius Dobzhansky, "Nothing in Biology Makes Sense Except in the Light of Evolution," *The American Biology Teacher* 35, no. 3 (1973): 125–129.
46. Michael Denton, *Evolution: A Theory in Crisis* (Bethesda, MD: Adler & Adler, 1986), 274–307.

47. See Christopher Howell, "Hope and the Ultimate Synthesis," *Public Orthodoxy*, January 13, 2022, https://publicorthodoxy.org/2022/01/13/hope-and-ultimate-synthesis/.
48. Teilhard de Chardin, *Phenomenon of Man*, 219.
49. Howell, "Hope and the Ultimate Synthesis."
50. Jitse M. van der Meer, "Theodosius Dobzhansky: Nothing in Evolution Makes Sense Except in the Light of Religion," in *Eminent Lives in Twentieth-Century Science & Religion*, 2nd ed., ed. Nicolaas A. Rupke (Frankfurt: Peter Lang, 2009), 110.
51. Quoted in Michael Shermer and Frank J. Sulloway, "Grand Old Man of Evolution: An Interview with Evolutionary Biologist Ernst Mayr," *Skeptic* 8 no. 1 (2000), republished in electronic format in 2004, https://www.skeptic.com/eskeptic/04-07-05/.
52. Howell, "Hope and the Ultimate Synthesis," cites Ayala's memoir for this point.
53. Christoph Schönborn, "Finding Design in Nature," *New York Times*, July 7, 2005, https://www.nytimes.com/2005/07/07/opinion/finding-design-in-nature.html
54. George Coyne, "God's Chance Creation," *Tablet*, June 8, 2005, https://users.ictp.it/~chelaf/Coyne.pdf.
55. Martin Hilbert, "Darwin's Divisions," *Touchstone*, June 2006, https://www.touchstonemag.com/archives/article.php?id=19-05-028-f.
56. "Kenneth Miller, 2017 St. Albert Award Winner," *Society of Catholic Scientists*, accessed October 3, 2024, https://catholicscientists.org/st-albert-award/kenneth-miller/.
57. Kenneth R. Miller and Joseph S. Levine. *Biology*: 4th ed. (Englewood Cliffs, NJ: Prentice Hall, 1998), 658.
58. Kenneth R. Miller, quoted in Michael O. Garvey, "Biologist Kenneth Miller to Receive Notre Dame's 2014 Laetare Medal," *Notre Dame News*, https://news.nd.edu/news/biologist-kenneth-miller-to-receive-notre-dames-2014-laetare-medal/.
59. Kenneth R. Miller, *Finding Darwin's God* (New York: Harper Collins, 1999), 272.
60. Miller, *Finding Darwin's God*, 274.
61. Michael Behe, *A Mousetrap for Darwin: Michael J. Behe Answers His Critics* (Seattle, WA: Discovery Institute Press, 2020), 67.
62. J. P. Moreland, "How Theistic Evolution Kicks Christianity out of the Plausibility Structure and Robs Christians of Confidence that the Bible is a Source of Knowledge," in *Theistic Evolution: A Scientific, Philosophical, and Theological Critique*, ed. J. P. Moreland (Wheaton, IL: Crossway, 2017), 644–645.
63. "Biologos Public Engagement: Inviting the Church and the World to See the Harmony between Science and Biblical Faith," Templeton Foundation, accessed April 30, 2021, https://www.templeton.org/grant/biologos-public-engagement-inviting-the-church-and-the-world-to-see-the-harmony-between-science-and-biblical-faith.
64. Chrisopher T. Baglow, *Faith, Science, & Reason: Theology on the Cutting Edge*, 2nd ed. (Downers Grove, IL: Midwest Theological Forum, 2019), 186.

65. Baglow, *Faith, Science, & Reason*, 189.
66. Baglow, *Faith, Science, & Reason*, 189.
67. Baglow, *Faith, Science, & Reason*, 189.
68. Chaberek, *Catholicism and Evolution*, 72–75, 93–140, 157–161.
69. Pius XII, *Humani Generis*, encyclical letter, Vatican, August 12, 1950, 37, https://www.vatican.va/content/pius-xii/en/encyclicals/documents/hf_p-xii_enc_12081950_humani-generis.html.
70. Paul VI, *Acta Apostolicae Sedis* 58 (Vatican 1966), 654, https://www.vatican.va/archive/aas/documents/AAS-58-1966-ocr.pdf
71. Kenneth W. Kemp, "Adam and Eve and Evolution," *Society of Catholic Scientists*, April 30, 2020, https://catholicscientists.org/articles/adam-eve-evolution/, quoted in Christopher Baglow, *Creation: A Catholic Guide to God and the Universe* (Notre Dame, IN: Ave Maria Press, 2021), 56.
72. Kenneth Miller, *The Human Instinct: How We Evolved to Have Reason, Consciousness, and Free Will* (New York: Simon and Schuster, 2018), 130–131.
73. Baglow, *Faith, Science, & Reason*, 261.
74. Baglow, *Creation*, 61–62.
75. See Edward T. Oakes, "Evolution and Original Sin," in *A Theology of Grace in Six Controversies* (Grand Rapids, MI: Eerdmans, 2016), 92–137.
76. *Catechism of the Catholic Church*, para. 389.
77. *Catechism of the Catholic Church*, para. 407.
78. *Catechism of the Catholic Church*, para. 390, emphasis in the original.
79. Nicanor Pier Giorgio Austriaco et al., *Thomistic Evolution: A Catholic Approach to Understanding Evolution in the Light of Faith* (Providence, RI: Cluny Media, 2016), 218.
80. Austriaco et al., *Thomistic Evolution*, 233.
81. Austriaco et al., *Thomistic Evolution*, 2nd ed. (2019), chap. 28.
82. See Stewart Candlish and George Wrisely, "Private Language," *Stanford Encyclopedia of Philosophy* (Fall 2019), https://plato.stanford.edu/entries/private-language/.

9. Converging and Convincing Arguments

1. In a letter to J. D. Hooker, dated March 29, 1863, Darwin writes: "I have long regretted that I truckled to public opinion & used Pentateuchal term of creation, by which I really meant 'appeared' by some wholly unknown process." *Darwin Correspondence Project*, Letter no. 4065, University of Cambridge, https://www.darwinproject.ac.uk/letter/?docId=letters/DCP-LETT-4065.xml.
2. Janet Browne, *Charles Darwin: The Power of Place* (New York: Alfred A. Knopf 2002), 318.
3. Stephen C. Meyer, *Darwin's Doubt: The Explosive Origin of Animal Life and the Case for Intelligent Design* (New York: Harper Collins, 2013), 204.

4. See Eric H. Anderson, "A Factory That Builds Factories That Builds Factories That …," in *Evolution and Intelligent Design in a Nutshell* (Seattle, WA: Discovery Institute Press, 2020), 65–86.
5. E. V. Koonin, *The Logic of Chance: The Nature and Origin of Biological Evolution* (Upper Saddle River, NJ: Pearson Education, 2012), 435.
6. See Denyse O'Leary, "The Passé Scientist," *Salvo* 56 (Spring 2021), 44-45, https://salvomag.com/article/salvo56/the-passe-scientist.
7. In November 2016, there was a meeting about evolution at the Royal Society. None of the proposed extensions of evolutionary theory—sometimes called "the third way," sometimes "the extended evolutionary synthesis"—could explain the origin of new information. See "Why the Royal Society Meeting Mattered, in a Nutshell," *Evolution News & Science Today*, December 5, 2016, https://evolutionnews.org/2016/12/why_the_royal_s/. Michael Behe has a chapter devoted to some of the proposals for an extended evolutionary synthesis, titled "Overextended," in *Darwin Devolves*, 115–137.
8. See *Darwin's Doubt*, chap. 6.
9. See Günter Bechly, "A Long Surrender," *Salvo* 62 (Fall 2022), https://salvomag.com/article/salvo62/a-long-surrender.
10. John Paul II, *Catechism of the Catholic Church* [1993], 2nd ed. (Washington DC: United States Catholic Conference, 2011), para. 31, https://www.usccb.org/sites/default/files/flipbooks/catechism/.
11. Psalm 18(19):1
12. Stephen Jay Gould, "A Visit to Dayton," in *Hen's Teeth and Horse's Toes* (New York: Norton 1983), 275.
13. Piotr Lenartowicz, *Ludy czy małpoludy? [People or Manapes?*, an English summary of Polish text] (Krakow, Poland: Jesuit University of Philosophy and Education Ignatianum, 2010), 260, https://lenartowicz.jezuici.pl/files/2011/09/54-SUMMARY1.pdf. The work initially circulated as a PDF, of which the URL here is a portion. In a later bound version of the book, the quotation appears on p. 383.

Appendix A. Some Notes on Transformism

1. Michael Chaberek, email to the author, June 14, 2021.
2. Jonathan McLatchie, "New Paper Argues That Variant Genetic Codes Are Best Explained by Common Design," *Evolution News & Science Today*, June 25, 2024, https://evolutionnews.org/2024/06/new-paper-argues-that-variant-genetic-codes-are-best-explained-by-common-design/. See also Winston Ewert, "On the Origin of the Codes: The Character and Distribution of Variant Genetic Codes Is Better Explained by Common Design than Evolutionary Theory," *BIO-Complexity* 2024, no. 1, https://bio-complexity.org/ojs/index.php/main/article/view/133.
3. Logan Paul Gage, "Can a Thomist Be a Darwinist?," chap. 10 in *God and Evolution*, ed. Jay Richards (Seattle: Discovery Institute, 2010),187–202, 201. The quoted passage from Thomas is from *Summa Contra Gentiles*, Part II, 99.

Appendix B. When Did Adam and Eve Live?

1. For an overview and discussion of recent mainstream scientific literature calling into question *Homo habilis*'s place in the genus *homo*, see Günter Bechly, "Fossil Friday: New Research Questions the Human Nature of *Homo habilis*," *Evolution News and Science Today*, January 19, 2024, https://evolutionnews.org/2024/01/fossil-friday-new-research-questions-the-human-nature-of-homo-habilis/. See also Casey Luskin, "Evolutionary Models of Palaeoanthropology, Genetics and Psychology Fail to Account for Human Origins: A review," in *Science and Faith in Dialogue*, eds. Frikkie Van Niekerk and Nico Vorster (Cape Town, South Africa: Aosis, 2022). For a discussion of *Homo naledi* and why it probably does not belong in *Homo*, see Casey Luskin, "Hominid Hype and *Homo naledi*: A Unique 'Species' of Unclear Evolutionary Importance," *Evolution News & Science Today*, September 20, 2015, https://evolutionnews.org/2015/09/hominid_hype_an/. For a discussion of how even mainstream scientists don't agree on how *Homo naledi* is related to other species, see "What Does It Mean to Be Human?," *Smithsonian Museum of Natural History*, January 3, 2024, https://humanorigins.si.edu/evidence/human-fossils/species/homo-naledi.

Figure Credits

Figure 2.1. The Tree of Life. "Haeckel arbol bn." Image by Ernst Haeckel, 1866, Wikimedia Commons. Public domain.

Figure 2.2. Bacterial flagellum. "Flagellum base diagram." Image by Mariana Ruiz Villarreal (Lady of Hats), 2007, Wikimedia Commons. Public domain.

Figure 2.3. Lime nail gall. "Eriophyes tilae tilae close up." Image by Roger Griffith (Rosser 1954), 2009, Wikimedia Commons. Public domain.

Figure 2.4. Haeckel's embryos. Image by G. J. Romanes (copied from Haeckel's original embryo drawings), 1892, Wikimedia Commons. Public domain.

Figure 2.5. The true "hourglass." Image by Jody F. Sjogren, 2000. Used with permission.

Figure 2.6. *Bathybius haeckeli*, the (incorrectly) supposed bridge between inanimate matter and life. "Bathybius haeckelii Haeckel 1870." Artist unknown, 1870, Wikimedia Commons. Public domain.

Figure 2.7. Schematic of the Miller-Urey experiment. "Miller-Urey experiment." Image by Carny, 2008, Wikimedia Commons. CC-BY-SA 3.0 license.

Figure 2.8. Amino acid chirality. "Chirality with hands." Image by NASA, Wikimedia Commons. Used as permitted.

Figure 3.2. Frontispiece of Bible Moralisée. "God the Geometer." Artist unknown, circa 1220–1230, Wikimedia Commons. Public domain.

Figure 5.1. Wasp pupae on caterpillar. "Cotesia congregata on Manduca sexta." Photograph by Beatriz Moisset, 2007, Wikimedia Commons. CC-BY-SA 4.0 license.

Figure 6.1. Frontispiece to Huxley's *Evidence as to Man's Place in Nature* (1863). "Huxley—Man's Place in Nature." Image by Benjamin Waterhouse Hawkins, 1863, Wikimedia Commons. Public domain.

Figure 6.2. Reconstruction of "Lucy." "Reconstruction of the fossil skeleton of 'Lucy' the *Australopithecus afarensis*." Image by User 120, 2007, Wikimedia Commons. Public domain. CC-BY-SA 3.0 license.

Figure 6.3. Mitochondrial Eve. "Mitochondrial Eve." Image by Michael Banate.

Figure 6.4. Chauvet Cave: Panel of the Lions. "Lions painting, Chauvet Cave (museum replica)." Photograph by HTI, 2009, Wikimedia Commons. Public domain.

Index

A

Abiogenesis, 83–90, 109, 275, 299
Adam and Eve, 16, 19, 24, 26, 31, 123–124, 129, 167–168, 169–171, 176, 179, 186–209, 212, 214, 238–240, 250, 252, 259, 264, 266–267, 269, 270, 282, 283–284, 285, 287, 289, 295–297
Agassiz, Louis, 216
Age of the earth, 100, 126–127, 205
Age of the universe, 22, 126, 128
Almécija, Sergio, 29, 177
Altizer, Thomas J. J., 253
Altruism, 167
Amino acids, 51–54, 67, 73, 86–88, 109, 110, 275, 306n34
Anthropic coincidences, 113, 260, 282
Ants, 64, 165
Aquinas, Thomas, 83, 98, 111, 124, 130–131, 142, 144–146, 149, 214–216, 238–240, 247–248, 268, 279, 281, 290, 292–293
Aristotle, 83, 94, 97–98, 110–111, 215, 221, 225, 240, 247–248, 290
Art, 25, 135–136, 174, 196–199, 201, 206, 221, 238, 284
Artificial intelligence, 25, 217–218, 231–238

Atheism (materialism), 7, 19, 27–28, 35, 55, 85, 98, 99, 113, 137, 139–140, 141, 153, 167, 209, 225, 242–246, 263, 274, 275, 276, 277, 280, 301
Augustine, 33, 124, 125, 214, 238, 266
Austriaco, Nicanor Pier Giorgio, 268–270, 286
Aveling, Edward, 244
Axe, Douglas, 54, 64, 275
Ayala, Francisco, 188–189, 256

B

Bacon, Francis, 97
Bacterial flagellum, 56–58, 258
Baggott, Peter, 122
Baglow, Christopher, 260–268, 286
Ball, Philip, 63–64
Bar-Hillel, Yehoshua, 234
Barlow, George, 68
Barr, Stephen, 236–237
Bathybius haeckelii, 84, 88
Beauregard, Mario, 225, 231
Behe, Michael, 21, 55–59, 68–69, 70–72, 100–101, 106–107, 114, 117, 118–121, 128, 138, 152–155, 258, 260, 277, 278, 279, 280, 290
Beijing man, 195, 250
Bell, Charles, 243

Benedict XVI (Ratzinger), 17, 18, 178, 252–253, 268
Bergson, Henri, 95, 246–247, 251
Berkeley, George, 142
Berlinski, David, 117
Bible, 22, 31, 117, 121, 122–129, 135–136, 139, 163, 203, 213–214, 218, 266, 278, 280–281, 289, 290–291, 300–301
Big Bang, 22, 72, 112, 113, 126, 128, 263, 292
Biogenetic law, 77–78
Biogeography, 75–76
BioLogos, 21, 189–190, 261
Black, Davidson, 250
Brain versus mind, 25, 39, 117, 138, 158, 223–237, 285
Breeding experiments, 38, 48–49, 291
Brent, James, 268
Browne, William, 243–245
Büchner, Ludwig, 244
Buggs, Richard, 190
Burgess shale, 44
Butterflies, 47, 62, 64

C

Cambrian explosion, 43–48, 110, 274
Campbell, Donald, 157–158
Carr, Bernard, 147
Carroll, Sean, 60, 120
Cavalli-Sforza, Luiga Luca, 200–201
Caverni, Raffaello, 263
Chaberek, Michael, 145, 249–251, 263–264, 268, 278, 290
Chen, J. Y., 115
Chimpsky, Nim, 219
Chinese-room experiment, 235–236
Chirality, 87
Chirikjian, Gregory, 238
Chomsky, Noam, 218, 219–223, 269–270

Cichlid fishes, 68–69, 72, 121, 308n69
Cladistics, 45–46
Classification systems, 45, 69, 74, 128, 20–205, 262, 289
Cobb, Matthew, 225–226
Coleridge, Samuel Taylor, 242
Collins, C. John, 124
Collins, Francis, 60, 261
Common descent, 24, 36, 61, 73–80, 153, 155, 186, 242, 278, 290, 291
Common design, 73–74
Convergent evolution, 74
Coppedge, David, 89–90
Cosmic beginning, 22, 55, 72, 112, 113, 126, 128, 263, 292
Cosmological constant, 112, 281
Cosmology, 22, 94, 111–113, 122, 149, 282
Council of Trent, 141–142, 287
Coyne, George, 25, 115–116, 257, 260, 286
Coyne, Jerry, 115–116, 138
Craig, William Lane, 31
Crick, Francis, 51, 89

D

Dark matter, 122
Darwin, Charles, religious beliefs, 18–19, 141, 242–246
Darwin, Emma (Wedgewood), 35, 246
Darwin, Erasmus, 34, 242–244
Darwin, George, 13, 15, 244
Darwin, Robert, 242–243
Darwinian racism, 165
Darwinism, 15–16, 21, 28–31, 34–72, 76–80, 81, 90, 97, 100, 134, 137, 153, 158, 274–277
Darwinism, metaphysical implications of, 13, 15, 16–17, 22, 25, 27, 28, 29, 31, 90,

112–113, 137, 153, 155–159, 162–163, 168, 177, 205, 213, 241–242, 245, 257, 260, 273, 283–287
Davenport, Thomas, 268
Davis, Mark, 183
Davison, John A., 49
Dawkins, Richard, 15, 21, 39, 54–55, 64, 91, 100, 104, 138, 163, 172–173, 260
Dawson, Charles, 249–250
Days of creation, 125, 291
De Bonald, Louis, 221–222
de Maillet, Benoît, 34
de Solages, Bruno, 247–248
De Waal, Frans, 183
Death, 19, 166–169, 172, 175, 205, 226–229, 239–240, 266, 267, 282
Deism, 113, 151, 246, 260, 263
Dembski, William, 23, 100–103, 107–110, 114
Demons, 132–133, 170
Dennett, Daniel, 113, 241–242
Denton, Michael, 60, 100, 111, 138, 220–222, 256, 279
Descartes, René, 94, 142
Design detection, 23, 101–113, 233
Design, bad, 161, 172–175
Designer, 139
Desmond, Adrian, 243
Determinism, 131–132
Devolution, 70–72, 121
Directed panspermia, 89
Dissent, stifling of, 20, 30, 114–116, 276, 278
DNA, 20, 51–63, 73–75, 80, 89, 101, 110–111, 134, 146, 185–192, 262, 269, 270, 275, 283
Dobzhansky, Theodosius, 25, 39, 49, 255–257, 260, 262
Doidge, Norman, 220
Dorlodot, Henri, 263

Dubois, Eugène, 179
Ducks, 165–166
Duhem, Pierre, 95

E

E. coli, 70–71
Eccles, John, 225
Eden, Murray, 52–53, 109
Egnor, Michael, 223–225
Ehmer, Anna Katharina, 229–230
Einstein, Albert, 97, 217
Eiseley, Loren, 85
Embryology, 77–80
Epigenetics, 61–64, 185, 276
Epistemology, 142, 157
Eschatology, 171–172, 175
Evangelization, 152, 154
Everett, Daniel, 219–220, 223
Evil and suffering, 13, 19, 24, 125, 151, 161–172, 176, 240, 242, 252–253, 265, 267, 282
Evo-devo, 60–61, 74, 276
Evolution, definition of, 35–37, 66–69
Evolutionary mechanism, edge of, 118–121
Evolutionary psychology, 138, 156–157, 158
Evolutionary theory, 15–16, 21, 28–31, 34–72, 76–80, 81, 90, 97, 100, 134, 137, 153, 158, 274–277
Ex nihilo, 73, 123, 127, 130–131, 152, 163, 238
Explanatory filter, 99–110
Eye, 14, 50–51, 60, 74, 161, 172–173, 174–175, 282

F

Fallen world, 151, 163, 167, 169, 170, 175, 176, 282
Feser, Edward, 110
Finches, 66–68, 70, 76, 121

Fine-tuning, 14, 46, 111–113, 114, 145, 151, 282
First human couple, 22, 26, 31, 123–124, 179, 186–209, 263, 267, 268, 283–284
Fisher, Ronald, 38–39
Fisher, Simon, 192
Flew, Antony, 93, 139
Fodor, Jerry, 21, 46, 91, 117, 138, 157–159
Foresight, 50, 54–55, 71, 280, 286, 292, 296, 306n34
Fossils, 28–29, 36, 41–48, 76, 113, 169, 178–185, 194 , 201, 202, 212, 213, 249–250, 274, 283–284, 291
Free agents, 131, 168
French revolution, 221, 243
Friedmann, Alexander, 126
Fruit flies, 49, 60, 69, 74, 255
Futuyma, Douglas, 173

G

Gage, Logan, 292
Gauger, Ann, 188–190
Genesis, 16, 33, 123–128, 169–170, 177, 189, 203–204, 208, 209, 213, 240, 262, 263, 266, 267, 278, 290
Genetic code, 20, 51, 61, 73–74, 291
Gilson, Étienne, 253–255
God of the gaps, 133–137, 282–283
God, existence of, 18, 24, 26, 116–117, 138–139, 141–159, 262, 281
God, nature of, 17, 19, 25, 40, 73, 102, 110, 113, 122–132, 143, 161–176, 240, 242, 246, 259–260, 263, 267, 279–280, 282
Gödel, Kurt, 232, 236–237
Goldschmidt, Richard, 49–50, 291
Gonzalez, Guillermo, 115, 281

Gould, Stephen Jay, 24, 40, 47, 49, 250, 259, 285
Grace, 26, 99, 140, 143, 170, 254
Graham, William, 216, 246
Grant, Peter and Rosemary, 67
Grant, Robert Edward, 243–245
Gray, Asa, 161, 216, 242
Greenberg, Joseph, 201
Greg, William, 244–245
Greyson, Bruce, 230
Grossman, Neal, 228–229

H

Haarsma, Deborah, 190
Haeckel, Ernst, 41–42, 76–79, 84, 88, 163
Haldane, John, 39, 69, 85
Happich, Friedrich, 229–230
Harold, Franklin, 58–59
Harvey, Ethel, 61
Harvey, William, 95
Hawking, Stephen, 238
Hedin, Eric, 115–116
Hegel, Georg Wilhelm Friedrich, 142
Hominin, definition of, 177–178
Homo erectus, 178–179, 191, 192, 193–194, 201–203, 208, 212, 283–284, 295
Homo sapiens, 22, 177–178, 179, 183, 190–193, 203–205, 283–284, 295
Hooker, Joseph Dalton, 34–35, 82–83
Hössjer, Ola, 190
Hox genes, 60–61
Hoyle, Fred, 55, 89
Hull, David, 40, 166–167
Human origins, 24, 28–29, 175–209, 211, 213, 250
Hume, David, 142
Hunter, Cornelius, 80, 162–163
Huxley, Julian, 39, 249, 251

Huxley, Thomas Henry, 15, 37, 39, 83–84, 88, 163, 178
Hybridization, 69, 276
Hydrothermal vents, 90
Hylomorphism, 215
Hypothetico-deductive method, 97

I
Illies, Joachim, 65
Image of God, 25, 129, 179, 193–194, 202, 207–209, 211–240, 284
Immediate cause, 102
Insect galls, 65–66
Instrumental cause, 102, 130
Intellect, 26, 214–218, 231–238
Intelligent design, history and definition of, 99–114
Irreducible complexity, 55–59, 101, 107, 260

J
Jaki, Stanley, 125, 153
Jaubert, Jacques, 198
John Paul II, 26, 33, 36, 80, 90, 150, 155–156, 213, 252, 286
Johnson, Phillip, 15–18, 40, 99–100, 122, 166
Johnston, George Sim, 15
"Junk" DNA, 59–60, 134

K
Kant, Immanuel, 142, 150, 158, 281
Keas, Michael, 36
Keller, Evelyn Fox, 217
Kemp, Kenneth W., 264–265
Kenyon, Dean, 86–88, 100, 115
Klein, Richard, 182
Koch, Christof, 227
Koonin, Eugene, 55, 109, 276

Ku, John Baptist, 268
Kurzweil, Ray, 237

L
Lagrange, Joseph-Louis, 96
Lamarck, Jean-Baptiste, 34
Lane, Nick, 89
Language, 24, 52, 104–106, 192, 200–20, 207, 212, 218–223, 234, 238, 269–270, 285
Laplace, Pierre-Simon, 131–132
Laplace's demon, 131–132
Larousse, Pierre, 81
Laughlin, Robert, 137
Leclerc, Georges-Louis, 34
Ledóchowski, Włodzimierz, 250
Leisola, Matti, 70
Lemaître, Georges, 126
Lenartowicz, Piotr, 204, 283, 287, 296
Lennox, John, 137
Lenski, Richard, 70–71
Leo XIII, 142
Leroy, Dalmace, 263
Levin, Michael, 63
Lewis, C. S., 14, 124, 170, 252
Lewontin, Richard, 53, 99, 129, 260, 283
Linnaean taxonomy, 289, 290
Liu, Renyi, 58
Locke, John, 142
Lönnig, Wolf-Ekkehard, 65
Lorber, John, 220
Loss-of-function mutations, 70–72, 121
Lucas, John, 236
Lucretius, 33–34
Luskin, Casey, 74–75, 183–184, 190
Lyell, Charles, 34–35, 40, 216

M
MacIntyre, Alasdair, 249
Macroevolution, 261

Malaria, 71, 118–119
Maritain, Jacques and Raïssa, 247
Materialism (see "Atheism")
Mathematics, 20, 38–39, 52–54, 94, 95–98, 100, 101–107, 109, 111, 131–132, 137, 144, 188–190, 236–237, 240, 275, 289
Matzke, Nicholas, 56–58
Mayr, Ernst, 30–31, 39, 53, 61, 184, 256
Mazur, Suzan, 81
McLatchie, Jonathan, 57–58
Medawar, Peter, 53, 248
Mendel, Gregor, 38, 258, 261
Mendelian genetics, 38, 261
Methodological materialism, 27, 93, 98, 113, 133, 134, 228, 230, 237, 283, 301
Meyer, Stephen, 36, 44–46, 117, 138, 277–278
Microevolution, 49, 192, 261, 284
Miller-Urey experiment, 86, 88
Miller, Kenneth, 25, 58, 79, 113, 119–121, 173, 258–260, 265, 267, 281–282, 286
Miller, Stanley, 85–86, 88
Mimetics, 47
Mind versus brain, 25, 39, 117, 138, 158, 223–237, 285
Minnich, Scott, 115
Miracles, 132, 154–155, 172, 281, 283
Mitochondrial Eve, 186–190
Modern synthesis (neo-Darwinism), 39, 261, 275, 276–277, 280
Monogenism, 22, 26, 31, 123–124, 179, 186–209, 263, 267, 268, 283–284
Moore, James, 243
Moran, Larry, 116–118
Moreland, J. P., 261
Morris, Simon Conway, 44
Most recent common ancestor (MRCA), 187–188

Mousetrap, 56, 106–107
Multiverse, 55, 109, 113, 147, 260, 276

N
Nadella, Satya, 238
Nagel, Thomas, 117, 121, 139–140, 277, 286
Nahm, Michael, 230
Natural law, 102, 156–159
Natural selection, 15–16, 21, 28–31, 34–72, 76–80, 81, 90, 97, 100, 134, 137, 153, 158, 274–277
Naturalism, 23, 25, 93, 133, 139, 377
Neanderthals, 179, 185, 190–193, 198, 199, 200, 205, 283
Near-death experiences, 25, 226–229, 285
Nelson, Paul, 53
Nephilim, 203, 204, 267
Neuroplasticity, 220–221, 285
Neutral evolution, 37
Newton, Isaac, 38, 72, 95–98, 132, 144
Nirenberg, Marshall, 51
Noah's flood, 126–127, 163, 204
Noble, Denis, 64
Nogar, Raymond, 248
Nominalism, 158, 215
Non-overlapping magisteria (NOMA), 22, 25, 135–137, 153, 261

O
O'Leary, Denyse, 225, 231
Ochman, Howard, 58
Ohno, Susumu, 59
Ontogeny, 78
Oparin, Alexander, 85
Origin of life, 34, 51, 55, 81–90, 93, 100, 117, 275, 276, 287
Original sin, 16, 23, 24, 27, 31, 167–168, 177, 203, 209,

238–240, 241, 250, 253, 263–267, 269–271, 285, 286, 287
Oxnard, C. E., 183

P
Pajaro Dunes meeting, 99–101
Paleontology, 47–48, 102, 103–104, 178–185, 206–207, 249–250, 274, 279, 283, 291, 295
Palevitz, Barry, 260
Paley, William, 155, 162, 172
Pallen, Mark, 56–57
Pasteur, Louis, 81, 83
Paul VI, 264
Penfield, Wilder, 25, 223–225, 285
Penrose, Roger, 112, 236–238
Permeability of creation, 129–133
Phyla, 45, 46, 72, 73, 290, 291
Phylogeny, 50, 75, 78
Physics, 20, 22, 72, 94, 95–99, 111–113, 117, 122–123, 126–132, 139, 145, 146, 151, 156, 159, 175, 215, 237, 260, 282, 283, 292
Piattelli-Palmarini, Massimo, 21, 46, 81, 91, 138
Piltdown man, 213, 249–250
Pius XII, 129, 203, 264, 296
Polygenism, 263–264
Popper, Karl, 97, 157–158
Poupard, Paul, 251–252
Premack, David, 222
Proteins, 51–54, 56–58, 59–60, 61, 63, 64, 67, 70, 73, 82, 86–88, 89, 108–110, 120, 137, 153, 186, 255, 262, 275, 306n34
Pseudogenes, 186

Q
Quantum mechanics, 32, 98, 122, 132
Quine, Willard Van Orman, 157

R
Racism, 164, 264–265
Ratzinger, Joseph, 17, 18, 178, 252–253, 268
Reason, 142, 149–151, 154, 167, 207, 218, 225, 235–237, 257, 269–270, 280, 296
Recapitulation theory, 77–78
Redemption, 26, 175
Redi, Francesco, 83, 95
Rehbock, Philip, 85
Reich, David, 190–193, 201
Richards, Jay, 36, 112, 281
Ridley, Mark, 90, 93
RNA world, 90
Robinson, Daniel, 225
Ross, Hugh, 171
Ruhlen, Merritt, 201

S
Sanger, Larry, 114–115
Sarmiento, Estaban, 182
Sarton, George, 95
Scadding, S. R., 80
Schönborn, Christoph, 257
Schützenberger, Marcel-Paul, 53
Science, definition of, 94–99
Science, rise of, 133–137
Scientism, 7, 20
Scruton, Roger, 150
Searle, John, 235–236
Seminal reasons, 33
Sermonti, Giuseppe, 33, 47, 60–62, 80–81, 129, 206–207, 241
Shedinger, Robert F., 30–31
Sickle cell anemia, 71
Simpson, George Gaylord, 177
Soul, 19, 25, 31, 39, 110–111, 123, 129–132, 134, 151, 155, 159, 167–168, 185, 193, 202, 205, 214–218, 221, 225–226, 231, 238, 239, 248, 258, 259, 266, 270, 282, 285, 286, 296, 300

Special creation, 53, 72, 121, 124, 163, 227, 289–291
Species, definition of, 67, 205
Specified complexity, 101, 104, 106–108, 134, 145, 146, 167, 168
Spinoza, Baruch, 142
Spiritual battle, 169–170, 176, 182
Spong, John Shelby, 26
Spontaneous generation, 81, 82–85
Sporns, Olaf, 225–226
Stark, Rodney, 221
Steno, Nicholas, 126
Sternberg, Richard, 115
Stick insects, 47
Struggle for survival, 165, 168, 208
Swamidass, Joshua, 190
Sykes, Rebecca Wragg, 198
Symbiogenesis, 75
Systems biology, 174

T

Taylor, Steven, 226–227
Teilhard de Chardin, Pierre, 17, 25, 246–263, 285
Teleology, 24, 139–159, 281
Templeton Foundation, 261
Terminal lucidity, 229–230
Termites, 62, 64
Terrace, Herbert, 219
Thaxton, Charles, 99
Theistic evolution, 21, 138, 147, 246–271
Theophilus, 149
Thomas, Neil, 29
Thomistic evolution, 268–271
Tour, James, 88–89
Transformism, 35–36, 289–293
Transitional forms, 22, 47, 97, 180–185, 192, 201, 202, 213, 249–250, 283, 291
Trasancos, Stacy, 152–154

Tree of life, 41–48, 74–75, 129, 182, 186, 262, 277
Trinity, 122, 213–214, 218, 238
Turing, Alan, 231–233
Tyndall, John, 83

U

Ulam, Stanislaw, 53
Umipeg, Vicki, 228
Uniformitarianism, 41
Urey, Harold, 85–88, 126
Ussher, James, 126, 204

V

Van der Meer, Jitse, 256
Van Lommel, Pim, 228
Vatican I, 141–142, 280
Vatican II, 18, 252
Venema, Dennis, 189–190
Verschuuren, Gerard M., 31
Voltaire, 13, 15, 29
Von Baer, Ernst, 77

W

Waddington, C. H., 53
Wallace, Alfred Russel, 34–35, 75, 217, 223, 274
Wang, Hao, 237
Watson, James, 51
Weinberg, Steven, 171
Wells, Jonathan, 172–173
West, John, 149, 155
Wetherington, Ronald, 182
Whitehead, Alfred North, 135
Wiedersheim, Ernst, 80
Wiker, Benjamin, 216, 242–243
Wikipedia, 114–115
Will, 238–240
Williams, George, 172–173
Wilson, Allan, 187
Wistar Institute, 53, 289
Witt, Jonathan, 70, 108
Wittgenstein, 270

Woit, Peter, 122
Wolfe, Tom, 219
Wood, Bernard, 182
Woodward, Arthur Smith, 249–250
Wright, Sewall, 39

X
Xavier, Joana, 89

Y
Y-chromosomal Adam, 186–190
Young-earth creationism (YEC), 116, 121–128, 205, 278, 289

Z
Zahm, John A., 263
Zallinger, Rudolph, 178

www.ingramcontent.com/pod-product-compliance
Lightning Source LLC
Chambersburg PA
CBHW020324170426
43200CB00006B/258